Springer Tracts in Modern Physics 90

Editor: G. Höhler
Associate Editor: E. A. Niekisch

Editorial Board: S. Flügge H. Haken J. Hamilton
H. Lehmann W. Paul

Springer Tracts in Modern Physics

66* **Quantum Statistics in Optics and Solid-State Physics**
With contributions by R. Graham, F. Haake

67* **Conformal Algebra in Space-Time and Operator Product Expansion**
By S. Ferrara, R. Gatto, A. F. Grillo

68* **Solid-State Physics** With contributions by D. Bäuerle, J. Behringer, D. Schmid

69* **Astrophysics** With contributions by G. Börner, J. Stewart, M. Walker

70* **Quantum Statistical Theories of Spontaneous Emission and their Relation to Other Approaches** By G. S. Agarwal

71 **Nuclear Physics** With contributions by J. S. Levinger, P. Singer, H. Überall

72 **Van der Waals Attraction:** Theory of Van der Waals Attraction By D. Langbein

73 **Excitons at High Density** Edited by H. Haken, S. Nikitine. With contributions by V. S. Bagaev, J. Biellmann, A. Bivas, J. Goll, M. Grosmann, J. B. Grun, H. Haken, E. Hanamura, R. Levy, H. Mahr, S. Nikitine, B. V. Novikov, E. I. Rashba, T. M. Rice, A. A. Rogachev, A. Schenzle, K. L. Shaklee

74 **Solid-State Physics** With contributions by G. Bauer, G. Borstel, H. J. Falge, A. Otto

75 **Light Scattering by Phonon-Polaritons** By R. Claus, L. Merten, J. Brandmüller

76 **Irreversible Properties of Type II Superconductors** By. H. Ullmaier

77 **Surface Physics** With contributions by K. Müller, P. Wißmann

78 **Solid-State Physics** With contributions by R. Dornhaus, G. Nimtz, W. Richter

79 **Elementary Particle Physics** With contributions by E. Paul, H. Rollnick, P. Stichel

80* **Neutron Physics** With contributions by L. Koester, A. Steyerl

81 **Point Defects in Metals I:** Introduction to the Theory 2nd Printing
By G. Leibfried, N. Breuer

82 **Electronic Structure of Noble Metals, and Polariton-Mediated Light Scattering**
With contributions by B. Bendow, B. Lengeler

83 **Electroproduction at Low Energy and Hadron Form Factors**
By E. Amaldi, S. P. Fubini, G. Furlan

84 **Collective Ion Acceleration** With contributions by C. L. Olson, U. Schumacher

85 **Solid Surface Physics** With contributions by J. Hölzl, F. K. Schulte, H. Wagner

86 **Electron-Positron Interactions** By B. H. Wiik, G. Wolf

87 **Point Defects in Metals II:** Dynamical Properties and Diffusion Controlled Reactions
With contributions by P. H. Dederichs, K. Schroeder, R. Zeller

88 **Excitation of Plasmons and Interband Transitions by Electrons** By H. Raether

89 Giant Resonance Phenomena in **Intermediate-Energy Nuclear Reactions**
By F. Cannata, H. Überall

90* **Jets of Hadrons** By W. Hofmann

91 **Structural Studies of Surfaces**
With contributions by K. Heinz, K. Müller, T. Engel, K. H. Rieder

92 **Single-Particle Rotations in Molecular Crystals** By W. Press

* denotes a volume which contains a Classified Index starting from Volume 36.

W. Hofmann

Jets of Hadrons

With 165 Figures

Springer-Verlag
Berlin Heidelberg New York 1981

Dr. Werner Hofmann
Institut für Physik, Universität Dortmund
D-4600 Dortmund, Fed. Rep. of Germany

Manuscripts for publication should be addressed to:
Gerhard Höhler
Institut für Theoretische Kernphysik der Universität Karlsruhe
Postfach 6380, D-7500 Karlsruhe 1, Fed. Rep. of Germany

Proofs and all correspondence concerning papers in the process of publication should be addressed to:
Ernst A. Niekisch
Haubourdinstrasse 6, D-5170 Jülich 1, Fed. Rep. of Germany

ISBN 3-540-10625-1 Springer-Verlag Berlin Heidelberg New York
ISBN 0-387-10625-1 Springer-Verlag New York Heidelberg Berlin

Library of Congress Cataloging in Publication Data. Hofmann, W. (Werner), 1952-. Jets of hadrons. Springer tracts in modern physics; 90). Bibliography: p. Includes index. 1. Hadrons. 2. Patrons. 3. Nuclear reactions. I. Title. II. Series. QC1.S797 vol. 90 [QC793.5.H322] 539s[539.7'216] 81-2276 AACR2

This work is subject to copyright. All rights are reserved, whether the whole or part of the material is concerned, specifically those of translation, reprinting, reuse of illustrations, broadcasting, reproduction by photocopying machine or similar means, and storage in data banks. Under § 54 of the German Copyright Law where copies are made for other than private use, a fee is payable to „Verwertungsgesellschaft Wort", Munich.

© by Springer-Verlag Berlin Heidelberg 1981
Printed in Germany

The use of registered names, trademarks, etc. in this publication does not imply, even in the absence of a specific statement, that such names are exempt from the relevant protective laws and regulations and therefore free for general use.

Offset printing and bookbinding: Brühlsche Universitätsdruckerei, Giessen
2153/3130 — 5 4 3 2 1 0

Preface

In the early seventies, progress in one of the major fields of elementary particle physics — the investigation of hadron-hadron interactions at high energies — seemed to stagnate. A large amount of experimental data was available, and many models were proposed to parametrize certain special features of data, but a common language to describe all phenomena in a unified manner, and to enable quantitative predictions was missing. However, a new era started with the investigation of those hadronic interactions where a large amount of transverse momentum is transferred to a single secondary particle. Many properties of these reactions could be understood in terms of the quark-parton model, which had already proven successful in the description of deep inelastic lepton-hadron scattering.

More recent developments succeeded to enlarge the range of applicability of the, now familiar, quark-parton concept to the bulk of hadron-hadron interactions.

Today it seems that the process of particle creation in almost all types of high energy reactions, like the decays of heavy resonances, e^+e^- annihilations into hadrons, lepton-hadron and hadron-hadron reactions, can be traced to one common origin: partons fragment under the influence of confinement forces into jets of hadrons.

The aim of this treatment is to discuss properties and phenomenology of hadron production in the different reactions, using the concept of jets of hadrons as a common guide line.

It is a pleasure for me to express appreciation to those who contributed to the existence and final form of this book, and especially to my colleagues in the CCHK/ACCDHW and DASP II collaborations. Particular thanks go to Prof. D. Wegener for innumerable valuable suggestions, many interesting discussions, steady encouragement and, last but not least, for careful reading of this manuscript. I acknowledge fruitful discussions with Prof. E. Reya and Prof. M. Glück concerning applications of QCD. I am also deeply indebted to my coworkers, Dr. A. Markees and Dr. J. Spengler for discussions and helpful comments. Finally, I thank Mrs. C. Strungat for her patient and perfect typing of various versions of the manuscript, as well as J. Huhn and Mrs. H. Bußmann who prepared the illustrations.

This work was supported by the "Bundesministerium für Forschung und Technologie".

Dortmund, January, 1981 *Werner Hofmann*

Contents

1. Introduction 1
2. Jets in e^+e^- Annihilations 4
   2.1 Evidence for Jets 5
   2.2 Inclusive Distributions 8
   2.3 Scaling Violations at High Energies 10
3. Jets in Longitudinal Phase Space Models 15
   3.1 The Uncorrelated Jet Model (UJM) 15
   3.2 The Approach to Scaling 19
   3.3 The UJM and Short Range Correlations 22
   3.4 Summary 24
4. Jets and Parton Models 25
   4.1 Jets from Quark Confinement 26
   4.2 Space-Time Development of Quark Jets 27
   4.3 An Algorithm for Simulation of Quark Jets 35
   4.4 Properties of Quark Jets 39
   4.5 Summary 45
5. Parton Jets and QCD 46
   5.1 Scale Breaking in QCD 47
   5.2 Preconfinement 55
   5.3 Production of Heavy Hadrons 59
   5.4 Transverse Momentum Structure of Parton Jets 61
   5.5 Gluon Jets 63
   5.6 Quantitative Test of QCD Predictions for Jets in e^+e^- Reactions .. 67
   5.7 Summary 68

6. The Fragmentation of Parton Systems 71
   6.1 Deep Inelastic Lepton-Nucleon Scattering 71
   6.2 Environmental Independence and Factorization 76
   6.3 Jet Universality 88
   6.4 Spectator Fragmentation 90
   6.5 The Quark Recombination Model (QRM) 95
   6.6 Dimensional Counting Rules 102
   6.7 The Three Gluon Decay of the Υ 108
   6.8 Summary 119

7. Jets in Hadron-Hadron Interactions with Particles of Large Transverse Momentum 121
   7.1 Parton-Parton Scattering 125
       7.1.1 The QCD Approach 130
       7.1.2 Parton Transverse Momentum 134
       7.1.3 Constituent Interchange Model (CIM) 136
   7.2 General Characteristics of High $p_\perp$ Events 138
       7.2.1 Particle and Beam Ratios 138
       7.2.2 Structure of Large $p_\perp$ Events 141
   7.3 The Jets of Large $p_\perp$ 148
       7.3.1 Longitudinal Distributions of Particles in Jets 148
       7.3.2 Transverse Properties of Jets and Five-Jet Final States 155
       7.3.3 Quantum-Number Correlations 159
   7.4 Spectator Fragmentation 166
       7.4.1 General Characteristics of the Spectator 167
       7.4.2 Quantum-Number Correlations 171
       7.4.3 Spectator Fragmentation Functions 172
       7.4.4 Particle Correlations in Spectator Jets 177
   7.5 Summary 179

8. Hadron-Hadron Interactions at Low $p_\perp$ 180
   8.1 Longitudinal Fragmentation Spectra 181

9. Summary 192

References 195
Subject Index 203
Classified Index 211

1. Introduction

The development of high energy physics during the last decade was characterized by the rapidly increasing energies of particle accelerators, thus allowing study of particle interactions at very high center of mass (cms) energies and momentum transfers. The step to four momentum transfers, which are large compared to the masses of the particles involved, lead to a new phenomenological model, the parton picture /1-4/. This model was strongly supported by the observation of the "jet" phenomenon in deep inelastic interactions /5-7/.

At high momentum transfers one observes that the emitted secondaries form bunches with a preferred axis, the jet axis (it is, however, difficult to give a general, model independent definition of a "jet", we shall postpone this to a later chapter). The first experimental evidence for the existence of jets, which originated from fragmenting partons, came independently from two reactions. First, in electron-positron annihilations at $\sqrt{s} > 5$ GeV the emitted particles tend to occupy a cylindrical volume of phase space, the longitudinal extension of which is large compared to its diameter /5/. The orientation of the cylinder, or jet axis, changes randomly from event to event. Secondly, jets have been observed in proton-proton interactions at ISR energies /6,7/. In a rare type of event, a particle of large transverse momentum with respect to the beam axis is produced /8-10/. Such events had been predicted as a consequence of a large-angle elastic scattering of partons /11/. A detailed analysis of the event structure revealed that these events contain two bunches of particles emitted at large angles, one of them includes the high $p_\perp$ particle, while the second jet opposite in azimuth compensates the transverse momentum of the first one /6,7/. The properties of jets seen in these two types of reactions turned out to be quite similar /7/.

It seems astonishing that the concept of jets has been introduced so late, in spite of the fact that in normal proton-proton interactions the produced secondaries are preferentially emitted along the direction of the incoming protons, thus in today's language, forming two jets.

There is, however, an essential difference: the jets seen in e^+e^- collisions and in high $p_\perp$ events are, at least on an event to event basis, not correlated with any obvious symmetry axis of the process, such as the direction of the primary particles. Since they are produced by large momentum transfers, with probe distances small compared to the typical hadron radii, these jets are intimately related to the parton structure of hadrons.

The jet structure visible in normal hadron-hadron collisions does not exhibit these particular features. It could be explained, e.g., in terms of multiperipheral models /12/, but in fact did not yield too much insight into the dynamics and systematics of strong interactions. Only recently, and using the parton picture as a guide line, physicists have started to see these phenomena as a manifestation of the parton structure of matter, and attempts to describe the observed particle spectra in terms of parton densities as measured in deep inelastic lepton-hadron interactions have been made /13,14/. Meanwhile, the production of jets resulting from the fragmentation of colored parton systems seems to be a common link between the various hadronic final states in normal hadron-hadron interactions, in lepton induced reactions, and in annihilation reactions /15/.

The aim of this work is to summarize our present knowledge on jets as a universal phenomenon in high energy physics and to investigate the dynamics of jet production and fragmentation.

This paper is organized as follows. In Chap.2 the experimental information on jets from e^+e^- annihilations is briefly summarized; a detailed discussion can be found in other volumes of this series /16/. Jets in e^+e^- annihilations are rather well known from both the experimental and the theoretical side, and are used as a reference for further discussions. In Chap.3 we try to describe the structure of jets by simple longitudinal phase space models, and the implications imposed by four-momentum conservation are studied. Chap.4 presents models for jet development in terms of the quark-parton language; the QCD corrections to the naive quark model are discussed in Chap.5. In Chap.6 we leave the "safe" ground of e^+e^- annihilations, and investigate the structure of the jets observed in lepton-hadron reactions and in decays of bound states of heavy quarks. Hadron-hadron interactions at large momentum transfers are considered in Chap.7. Phenomenological models for the fragmentation of the multiquark systems involved will be compared to recent data. In Chap.8 extensions of the quark-parton jet concept to hadronic interactions at low momentum transfers are studied. Finally a brief summary is given.

This work is intended to give an introductory review to the phenomenon of jets in hadronic final states, it contains therefore quite a lot of "old" physics, which seems to be relevant for the understanding of jets. Since the major aim is to discuss physical concepts and models, no attempt has been made to give an exhaustive compilation of data and references, only those actually used in the discussion are quoted. We take this opportunity to apologize to all those authors whose important contributions to the field are not quoted because of ignorance or lack of space.

2. Jets in e^+e^- Annihilations

When the first electron-position storage rings in the GeV range came into operation, one of the surprising results was that over a wide range of energies the ratio R of the hadronic cross section to the total cross section for muon pair production is approximately constant. This implies a pointlike coupling of the virtual photon to the hadronic final state. A natural explanation of this phenomenon is given by the quark-parton model /1-4/: the virtual photon creates a quark-antiquark pair. According to the postulates of the quark model, these quarks fragment into observable hadrons with unit probability. Quark creations and decays are governed by violently different time scales and thus may be treated independently. As the photon-quark coupling is determined by the square of the quark charge, one obtains in the quark model /17-19/

$$R = \frac{\sigma(e^+e^- \to \text{hadrons})}{\sigma(e^+e^- \to \mu^+\mu^-)} = \sum_{\substack{\text{quark flavors and colors}}} q_i^2 + \text{QCD corrections of } O(\alpha_s/\pi). \qquad (2.1)$$

The sum is extended over all quark species above threshold. Since the four momentum of the virtual photon is timelike there are steps in R due to the production of new heavy quark flavors. Close to these thresholds the relative velocity of the quarks is small, and nonperturbative effects like resonance formation dominate. Therefore (2.1) does not hold in the vicinity of thresholds. In Fig.2.1 the measured values of R are compared with (2.1), including the known thresholds for charmed, and bottom quarks. As in all following figures, contributions from heavy-lepton decays have been subtracted, or are irrelevant. The approximate agreement of experimental data with theory supports the assumptions that quarks dress into hadrons with unit probability and that this mechanism is sufficiently soft not to interfere with the hard production process.

In a soft fragmentation process, the final state hadrons are expected to have small transverse momenta with respect to the quark axis of flight. Noting further that the mean hadron multiplicity in e^+e^- annihilations (Fig.2.2) rises slower

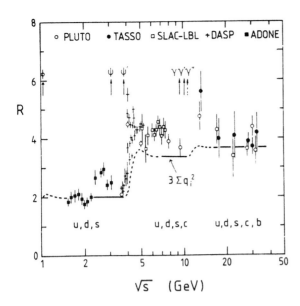

Fig.2.1. The ratio R of hadronic to µµ cross sections as measured by the γγ group at ADONE /21/, SLAC-LBL /22/, PLUTO /22/, DASP /20/ and TASSO /24/. Contributions from heavy-lepton decays subtracted

than the cms energy, which is equivalent to an increase of the mean hadron momenta with s, one expects that the final state particles form two jets around the directions of the two quarks.

2.1 Evidence for Jets

Various approaches are possible to prove the two-jet structure of the hadronic final state. At very high energies, the mean momentum parallel to the jet axis should be large compared to the transverse momentum for typical fragments, and one may hope to "see" a jet structure simply by inspection of the events. This is in fact true at PETRA energies, as can be seen from Fig.2.3.

However, for further investigations it is necessary to have a quantitative measure for the jetiness of hadronic final states. This can be achieved by two different methods. First, a number measuring the jetiness can be compiled using the four momenta of all produced particles (or, as a first approximation, of all detected particles) /28-31/. Secondly, one can study inclusive multiparticle correlations. One of the most commonly used, being experimentally convenient quantities of the first type, is the sphericity $\tilde{S}$ /32/

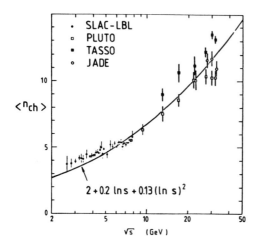

Fig.2.2. Mean charged multiplicity in e^+e^- annihilations vs $\sqrt{s}$. Data from SLAC-LBL /25/, PLUTO /26/, JADE /27/, and TASSO /24/. Contributions from heavy-lepton decays are subtracted, i.e., are irrelevant. The errors shown for the high-energy data refer to statistical errors only, systematic errors are typically ± 1 unit

$$\tilde{S} = \frac{3}{2} \min \frac{\Sigma p_{\perp i}^2}{\Sigma p_i^2} \tag{2.2}$$

where $p_{\perp i}$ is the transverse momentum of the ith particle with respect to an axis chosen to minimize $\tilde{S}$.

The expected behavior of $\tilde{S}$ is as follows: at low energies and multiplicities, momentum conservation requires the produced particles to move back to back, and $\tilde{S}$ will be small. At higher energies, phase space models predict the particle distribution to be more and more spherical, and $\tilde{S}$ rises with energy to the asymptotic value 1. In contrast, jet models with limited transverse momentum with respect to the jet or sphericity axis predict asymptotically

$$\tilde{S} \simeq \frac{3}{2} \frac{\langle p_\perp^2 \rangle}{\langle p^2 \rangle} \sim \left(\frac{\langle n \rangle}{\sqrt{s}}\right)^2 \langle p_\perp^2 \rangle \tag{2.3}$$

With increasing energy, $\tilde{S}$ increases until $\langle p^2 \rangle \gg \langle p_\perp^2 \rangle$, and then decreases according to (2.3). First evidence for a jet structure, seen as a decrease of $\tilde{S}$ as the energy increases, came from the SLAC-LBL magnetic detector at SPEAR, at cms ener-

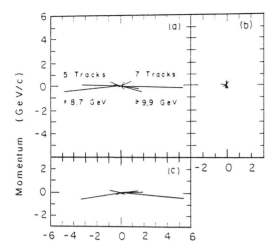

Fig.2.3. Momentum space representation of a typical two-jet event /27/, shown in three orthogonal projections

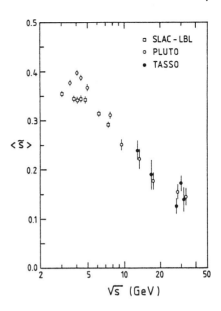

Fig.2.4. Mean sphericity as a function of $\sqrt{s}$. Data from /5,24,26/

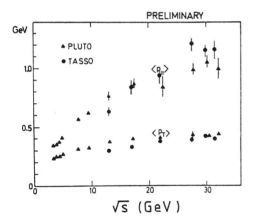

Fig.2.5. Mean momentum components parallel and perpendicular to the jet axis defined by thrust /24,26/

gies above 5 GeV /5/. Today's knowledge on $\tilde{S}$ is summarized in Fig.2.4. The data clearly favor the jet-like production of particles.

Besides exclusive jet measures, information on a jet-like structure of the final state can be obtained from inclusive spectra, and from correlation data.

Figure 2.5 shows the mean momentum, parallel ($p_\parallel$) and transverse ($p_\perp$) with respect to the jet axis, found by maximizing the sum of the momentum components parallel to this "thrust" axis. As expected for jets, the mean transverse momentum is roughly independent of the jet energy, and of the same order of magnitude as observed in hadron-hadron interactions. The numerical values should be taken with some care, however, since they depend on the method used to define the jet axis. In contrast to the mean transverse momentum, the mean longitudinal momentum increases continuously with energy. Of course due to the selection of the jet axis, $<p_\shortparallel>$ will always be slightly larger than $<p_\perp>$. This selection bias should be strongest at low energies, or multiplicities. The difference between $<p_\shortparallel>$ and $<p_\perp>$, which increases with energy, is a clear signature for jets.

2.2 Inclusive Distributions

We start with a brief description of the formalism for inclusive production /33,34/

$$e^+e^- \to h+x$$

The hadron h is characterized by its mass m and its four momentum $p = (E,\underline{p})$.

The cross section for h production (when summing over the polarization of h, e^+, and e^-) is described by two structure functions, corresponding to the longitudinal and transverse polarization components of the virtual photon

$$\frac{d^2\sigma}{dx_R d\Omega} \simeq \frac{\alpha}{s} x_R (-F_1^h + \frac{1}{4} x_R F_2^h \sin^2\theta) \tag{2.4}$$

with $x_R = 2E/\sqrt{s}$. The structure functions F_1^h and F_2^h depend on E, s, and on the type of h. At high energies, where m is small, one gets the scaling limit for particle production via intermediate spin 1/2 partons:

$$F_1^h(E,s,m) \to F_1^h(x_R) = \frac{1}{2} x_R F_2^h(x_R) \tag{2.5}$$

$sd\sigma/dx_R$ is then expected to scale as /33,34/

$$sd\sigma/dx_R \to (1-x_R)^n/x_R \tag{2.6}$$

The term 1/x accounts for the rise of the mean multiplicity with s. In the limit $E \gg m$, x_R can be replaced by $x = 2|\underline{p}|/\sqrt{s}$. Experimental results for $d\sigma/dx$ are shown in Figs. 2.6 and 2.7 for the energy range $3.0 \le \sqrt{s} \le 7.4$ and $5.0 \le \sqrt{s} \le 31.6$ GeV, respectively. In the low-energy region data from the SLAC-LBL /35/ detector has been chosen as a representative sample, the high-energy results come from the

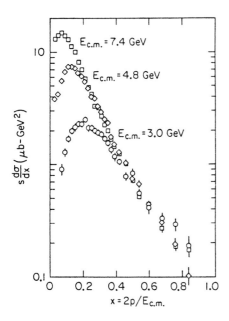
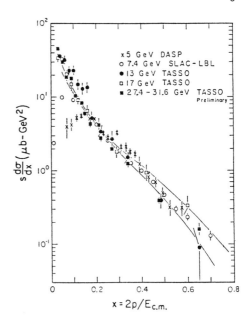

Fig.2.6. Inclusive cross section $sd\sigma/dx$ vs s. Data from the SLAC-LBL detector /44/

Fig.2.7. Inclusive cross section $sd\sigma/dx$ from DASP /36/, SLAC-LBL /35/, and TASSO /24/. The curves $(1/x)(1-x)^2$ and $(1/x)(1-x)^3$ shown for comparison are normalized at $x = 0.2$

TASSO detector at PETRA /24/. With the exceptions of the threshold region $x \simeq 0 (m/\sqrt{s})$ and eventually except the very lowest energy, all spectra scale in x within the accuracy of the measurements.

For further studies and to enable a comparison with the jets observed in hadron-hadron collisions, it is convenient to use quantities referring to the jet axis, like the rapidity $y = (1/2)\ln[(E+p_{||})/(E-p_{||})]$, or the transverse momentum $p_\perp$. These quantities are, of course, somewhat sensitive to the way the jet axis is defined, the influence being strongest for very slow $[(|p| \simeq 0\ (300\ \text{MeV})]$ and very fast $(x \to 1)$ particles.

Figure 2.8 shows the rapidity distribution per event, $(1/\sigma)(d\sigma/dy)$ as measured by the SLAC-LBL /22/ and TASSO /24/ detectors, with respect to the sphericity (SLAC-LBL) and thrust axis (TASSO). For the SLAC-LBL data, a fast particle $(x > 0.3)$ in one of the jets is required in order to allow a reliable determination of a jet axis already at energies as low as 5 GeV. The density shown is twice the density observed in the hemisphere opposite to the fast particle. As shown in /22/, this method introduces nearly no bias.

The rapidity distributions show the following features:

- the length of the rapidity interval populated increases as ln(s)
- at high energies, the rapidity distribution develops a plateau around y = 0
- the height of the plateau seems to increase slightly with increasing cms energy.

For comparison, the rapidity distribution observed for mesons produced in proton-proton interactions at $\sqrt{s}$ = 31 GeV /37/ is included in Fig.2.8. It is in surprisingly good agreement with the high-energy TASSO data, taking into account that in proton-proton interactions a sizeable fraction of energy is carried off by leading nucleons, so that as far as meson production is concerned, the proton-proton data at $\sqrt{s}$ = 31 GeV should be compared with e^+e^- annihilations at $\sqrt{s} \simeq$ 15-20 GeV. There are indications that the height of the plateau is slightly larger for the e^+e^- data. Furthermore, the distributions seem to develop a hole at y = 0. The second effect may be due to biases introduced by the definition of the jet axis.

Figure 2.9 shows the distribution of transverse momenta at $\sqrt{s} \simeq$ 7 /22/, $\sqrt{s} \simeq$ 13-17 and $\sqrt{s} \simeq$ 30 GeV /24/. As a reference, the $p_\perp$ distribution of pions produced in proton-proton collision at $\sqrt{s} \simeq$ 14-20 GeV is shown /38/. Taking into account that at the lowest energy transverse momenta above 1 GeV/c are damped due to energy and momentum conservation, all data in the range $\sqrt{s} \simeq$ 7 GeV to $\sqrt{s} \simeq$ 20 GeV agree. Again there is no severe difference between hadrons produced in proton-proton and in electron-positron collisions. However, the $p_\perp$ distribution at $\sqrt{s} \simeq$ 30 GeV develops a strong tail towards higher $p_\perp$'s, starting at $p_\perp^2 \simeq$ 0.5 GeV^2. There is no correspondent feature observed in proton-proton interactions; in the $p_\perp$ range studied here the slope of the $p_\perp$ distribution is nearly independent of $\sqrt{s}$ for $\sqrt{s}$ > 15 GeV. We shall return to these phenomena in Sect.2.3.

The dependence of the mean transverse momentum on $x_\parallel$ is shown in Fig.2.10 for two energies. Note that the shape of this curve is sensitive to the way the jet axis is defined, systematic changes at low and high $x_\parallel$ may be as large as 20%-30%. Like the corresponding proton-proton data, the distributions show a pronounced seagull effect at $x_\parallel \simeq$ 0.

2.3 Scaling Violations at High Energies

In Sect.2.2 we found that in the energy range 5 < $\sqrt{s}$ < 20 GeV the inclusive particle distributions in jets scale in x_R and have a fixed, limited $p_\perp$ with respect to the jet axis, the deviations from universal distributions at low x_R and at large $p_\perp$ being induced by phase space effects. At $\sqrt{s} \simeq$ 30 GeV, however, the $p_\perp$ distribution changes qualitatively, it develops a tail towards larger $p_\perp$, which must have

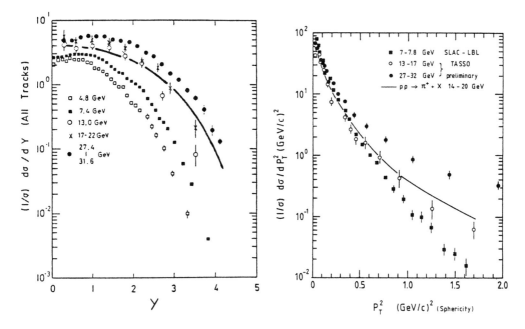

Fig.2.8 Fig.2.9

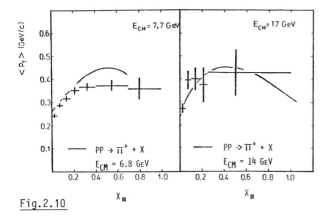

Fig.2.10

Fig.2.8. Rapidity distributions for charged particles assuming $m = m_\pi$, measured at $\sqrt{s} = 4.8, 7.4$ /22/ and at 13,17,27.6-31.6 GeV /24/. The full line shows the distribution of secondaries in inelastic pp interactions at $\sqrt{s} = 31$ GeV /37/

Fig.2.9. Distribution of transverse momenta with respect to the sphericity axis /22,24/. The SLAC-LBL and the proton-proton data /38/ are normalized arbitrarily

Fig.2.10. Seagull plot $<p_\perp>$ vs $x_\shortparallel$ from charged tracks /23/, compared to data from proton-proton interactions $\sqrt{s} = 6.8$ and 14 GeV /38-40/

a dynamical origin (Fig.2.9). In the following we shall investigate this effect in more detail. For the discussion, we will mainly use results from TASSO /24/; similar results have been reported from the PLUTO /41/, MARK-J /42/ and JADE /43/ collaborations. The widening of the $p_\perp$ distribution can have the following origins:

- the production of a new quark flavor;
- the $p_\perp$ distribution for quark fragmentation is energy dependent, the average $p_\perp$ grows as s increases;
- as in hadron-hadron interactions /6-11/, single hard scattering processes dominate the cross section at large $p_\perp$ and s.

Candidates for such processes are, e.g., the emission of hard gluon bremsstrahlung under large angles by the primary quark /44,45/ or the production of meson resonances with large transverse momenta by constituent interchange mechanisms /46/.

The first possibility can be immediately ruled out by other data /24/. The other two possibilities can be discriminated by examining the phase space structure of the events: an increase of the fragmentation-$p_\perp$ with energy still yields particle distributions which are symmetric in azimuth with respect to the jet axis. Hard parton processes produce a third jet, either by gluon fragmentation in quark-gluon models or by the decay of the excited meson in the CIM model /46/. In most cases this jet will be more or less aligned with one of the initial jets, and the cross section of the cylinder will be deformed into an ellipsoid; sometimes, a planar three-jet structure will be recognizable.

In search for planar events, the TASSO collaboration assigned an event plane in such a way as to minimize the sum of the momentum components squared out of this plane /47/. Obviously, the jet axis as defined via the minimum sphericity is contained in that plane. Figure 2.11 shows the mean transverse momentum squared in the event plane, $<p_\perp^2>_{IN}$, and normal to the event plane, $<p_\perp^2>_{OUT}$, per event at $\sqrt{s}$ from 13 to 17 GeV and at $\sqrt{s}$ around 30 GeV /24/. By definition of the event plane, $<p_\perp^2>_{IN}$ is larger than $<p_\perp^2>_{OUT}$. The comparison of the low- and high-energy data proves that this bias is not able to explain the tail at $<p_\perp^2>_{IN}$ in the 30 GeV data. In fact, models with a fragmentation symmetrically in azimuth are not able to describe this tail consistently. This observation is very much in favor of the last hypothesis. In fact, events sitting in the tail of the $<p_\perp^2>_{IN}$ distribution show a three-jet signature. Fig.2.12 gives a "typical" example.

The planar events were analyzed as three-jet events. The distribution of the transverse momentum with respect to the axis was measured for each jet and was compared to the inclusive $p_\perp$ distribution of two-jet events at $\sqrt{s}$ = 12 GeV (Fig.2.13). Obviously the events which lead to large $<p_\perp>$ values when treated as two-jet events have the canonical $<p_\perp>$ value of ~0.3 GeV when analyzed as three-jet events.

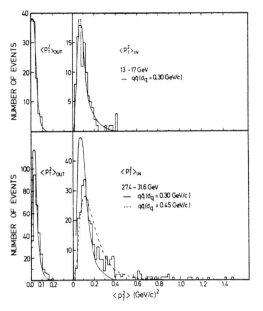

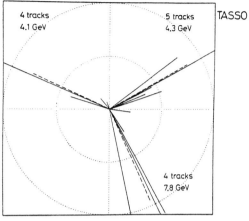

Fig.2.11

Fig.2.12

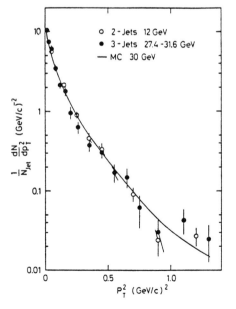

Fig.2.11. Mean transverse momentum squared in and perpendicular to the event plane per event. The full and dotted lines are predictions of jet models using particle distributions symetrically in azimuth /24/

Fig.2.12. A three-jet event projected into the event plane /24/

Fig.2.13. Observed transverse momentum distribution for noncollinear planar events at $\sqrt{s} \cong 30$ GeV, when analyzed as three jet events, compared to the inclusive $p_\perp$ distribution of events at $\sqrt{s} \cong 12$ GeV, analyzed as two-jet events /24/

Fig.2.13

Does the hypothesis of a hard-bremsstrahlung process explain the sudden onset of the effect when going from s = 17 to s ≅ 30? The probability for emission of a hard field quantum is given by /48/

$$w \cong \text{const.} \frac{\alpha_S}{\pi} \ln(s/\mu^2) \qquad (2.7)$$

where α_S is the strong coupling constant and μ is a cut off parameter of the order 1 GeV. Rather slowly does w vary with energy. Two further circumstances are relevant however. First, a third jet emitted at large angles can be recognized as soon as the mean longitudinal momentum of fragments is larger than $<p_\perp>$. Since $<p_{||}>$ is roughly proportional to the jet energy, this criterion introduces a much faster variation in the fraction of events which can be identified as multijet events. Secondly, there are two competing mechanisms governing the energy loss of a fast quark: hard bremsstrahlung and soft hadronization (although we shall see in Chap.5 that a strict distinction of the two mechanisms is not justified). Their relative importance depends on the ratio of the corresponding time scales. As we shall note in Chap.4, the fragmentation time of the jet is of the order

$$\tau_{frag.} \sim \frac{\sqrt{s}}{m_0^2} \qquad (2.8)$$

with m_0 being a typical hadronic mass scale. On the other hand, the partial lifetime for emission of a quantum carrying a fraction z of the quarks momentum at an angle θ is given by /48/

$$\tau_{brems.} \sim \frac{1}{\sqrt{s}\, z(1-z)\sin^2\theta/2} \qquad (2.9)$$

From (2.8) and (2.9) follows that the relative importance of hard processes increases proportional to s. This explains the fast transition from the hadronisation-dominated regime to the region where hard interactions are obvious.

A unique identification of the hard process as gluon bremsstrahlung is difficult at the present quality of data; however, the main features of the data are consistent with this interpretation /49,50/.

A more detailed discussion on hard processes in jet fragmentation will follow in Chap.5.

3. Jets in Longitudinal Phase Space Models

As we have seen in the last chapter, the physics of jets at small energies, up to $\sqrt{s} \simeq 10\text{-}20$ GeV, is dominated by phase space effects due to the nonzero transverse momenta and particle masses. To arrive at a quantitative description of the limitations imposed by pure four-momentum conservation and to separate jet dynamics from kinematics, it is useful first to study a simple phase space model for jet production.

This section is organized as follows. We shall present the uncorrelated jet model for e^+e^- annihilations, discuss the physical meaning of the parameters involved, and give the asymptotic behavior of the model. Next, scaling variables and the approach to the scaling limit will be discussed. In the remainder of the chapter, possible generalizations of the model are presented, e.g., the inclusion of resonance production and the use of more sophisticated matrix elements.

3.1 The Uncorrelated Jet Model (UJM)

In the UJM the fully exclusive decay probability of a virtual photon of the four momentum $Q = (\sqrt{s}, 0)$ into N particles is assumed to factorize as /51,52/

$$\Gamma_N \sim \left[\prod_{i=1}^{N} \frac{d^3 p_i}{E_i} f(p_\parallel^i, p_\perp^i) \right] \delta^4(\sum p_i - Q) \tag{3.1}$$

neglecting spins and photon polarization. The four momentum of the ith secondary is $p_i = (E_i, \underline{p}_i)$, which for simplicity will be assumed to be a chargeless pion. To arrive at final state jets, the invariant momentum space element $d^3 p_i/E_i$ /53/ is weighted with a matrix element f depending on the longitudinal and transverse momenta with respect to the jet axis defined by a unit vector $\hat{\underline{e}}$

$$p_\parallel = \underline{p} \cdot \hat{\underline{e}}; \quad p_\perp = |\underline{p} \times \hat{\underline{e}}| \tag{3.2}$$

The δ-function in (3.1) accounts for energy-momentum conservation. As the n-particle matrix element factorizes into n independent probabilities, the model contains no dynamical correlations. In the following we shall adopt the simplest choice for f, generating a transverse-momentum cutoff

$$f(p_\shortparallel, p_\perp) \sim \exp(-\lambda p_\perp) \tag{3.3}$$

Such models have been discussed by many authors; we shall follow the presentation of /54/.

In terms of the momentum space volume (the grand partition function)

$$\Omega(\hat{e},Q) = \sum_{N=2}^{\infty} \frac{\kappa^N}{N!} \Gamma_N \tag{3.4}$$

the inclusive single-particle spectrum in the photon rest frame, normalized to the total cross section σ, is given by

$$(E/\sigma)(d^3\sigma/dp^3) = \kappa \exp(-\lambda p_\perp) \Omega(\hat{e}, Q-p)/\Omega(\hat{e},Q) \tag{3.5}$$

with the trival sum rules ($x_R = 2E/\sqrt{s}$)

$$\frac{1}{\sigma} \int dx_R dp_\perp^2 \frac{d^3\sigma}{dx_R dp_\perp^2} = \langle N \rangle; \quad \frac{1}{\sigma} \int dx_R dp_\perp^2 \, x_R \frac{d^3\sigma}{dx_R dp_\perp^2} = 2 \tag{3.6}$$

The physical significance of the parameters κ and λ is evident: λ^{-1} determines the transverse jet width and κ characterizes the multiplicity distribution, an increase in κ giving a higher weight to larger particle numbers. Asymptotically, one obtains

$$\frac{E}{\sigma} \frac{d^3\sigma}{dp^3} = \frac{1}{\sigma} \frac{1}{\pi} \frac{d^2\sigma}{dy\, dp_\perp^2} \underset{\substack{S \to \infty \\ y \text{ fixed}}}{=} \kappa \exp(-\lambda p_\perp) \tag{3.7}$$

Consequently, the particle density per unit of rapidity is

$$\frac{1}{\sigma} \frac{d\sigma}{dy} = \frac{\kappa \pi}{\lambda^2} \underset{\substack{S \to \infty \\ y \text{ fixed}}}{=} \tilde{\kappa} \tag{3.8}$$

leading to a mean multiplicity

$$<N> = \tilde{\kappa}\ln(S) \quad S \to \infty \tag{3.9}$$

The asymptotic behavior of the inclusive x_R distribution is given by (3.1) and (3.5) if we note that $\Omega(\hat{\underline{e}},Q-p)$ is a function of the particles transverse mass $m_\perp$ and of the longitudinal missing mass M_L

$$M_L = [(\sqrt{s}-E)^2 - (\hat{\underline{e}}\underline{p})^2]^{1/2} = \sqrt{s}\left(1-x_R+\frac{m_\perp^2}{S}\right)^{1/2} \tag{3.10}$$

with the asymptotic limit

$$\Omega(\hat{\underline{e}},Q-p) \sim \frac{M_L^{2\tilde{\kappa}-2}}{\ln(M_L/m_\perp)}[1+O(\ln^{-1}(M_L/m_\perp)] \tag{3.11}$$

$$m_\perp \ll M_L, \sqrt{s}$$

For $1-x \gg m_\perp^2/s$ and large s, this amounts to scaling of

$$\frac{E}{\sigma}\frac{d^3\sigma}{dp^3} \cong \kappa[\exp(-\lambda p_\perp)](1-x_R)^{\tilde{\kappa}-1} \quad \text{or} \quad \frac{1}{\sigma}\frac{d\sigma}{dx_R} \cong \frac{2\tilde{\kappa}}{x_R}(1-x_R)^{\tilde{\kappa}-1} \tag{3.12}$$

in the photon rest system.

Up to now, we used the simplest matrix element which is able to describe the transverse-momentum cutoff, $f(p_{\shortparallel},p_\perp,S) \sim \exp(-\lambda p_\perp)$. More sophisticated UJM-matrix elements have been analyzed in /55/, with a rather astonishing result.

In the high-energy scaling limit the most general form of f is

$$f(p_{\shortparallel},p_\perp,S) \underset{S\to\infty}{=} f(x,p_\perp) \tag{3.13}$$

Define

$$f_{\shortparallel}(x) = \int_0^\infty dp_\perp^2\, f(x,p_\perp) \tag{3.14}$$

The inclusive cross section is now given by /55/

$$\frac{1}{\sigma}x_R\frac{d\sigma}{dx_R} \cong \frac{1}{\sigma}x\frac{d\sigma}{dx} = \tilde{\kappa}g(x)(1-x)^{\tilde{\kappa}-1} \tag{3.15}$$
$$S\to\infty$$

In general, g(x) is a rather flat function of x, even if $f_{\shortparallel}(x)$ has a large variation in x. Physically this is due to the fact that if we try to change the inclusive distribution by changing $f_{\shortparallel}(x)$, let's say give it a sharper decrease in x, we give less freedom to the longitudinal momentum of the measured particle. But large momenta of the other particles are suppressed as well so that the damping due to the four-momentum conservation will be less and so will compensate for the change in $f_{\shortparallel}(x)$. This is most drastically shown by noting that $g(x)$ is invariant under the transformation

$$f_{\shortparallel}(x) \to \exp(-\alpha x)\, f_{\shortparallel}(x);$$

especially we have

$$g(x) = 1 \quad \text{if} \quad f_{\shortparallel}(x) = \exp(-\alpha x) \tag{3.16}$$

So, one sees that independent of $f(x,p_\perp)$, $(1/\sigma)(xd\sigma/dx)$ will always be a function very similar to $\kappa(1-x)^{\kappa-1}$. Accordingly, the two particle distributions turn out to be

$$\frac{1}{\sigma} x_1 x_2 \frac{d^2\sigma}{dx_1 dx_2} \underset{S\to\infty}{=} \begin{cases} \tilde{\kappa}^2 (1-x_1-x_2)^{\tilde{\kappa}-1} \theta(1-x_1-x_2) \\ \text{if 1,2 are in the same hemisphere} \\ \tilde{\kappa}^2 (1-x_1)^{\tilde{\kappa}-1} (1-x_2)^{\tilde{\kappa}-1} \\ \text{if 1,2 are in opposite hemispheres} \end{cases} \tag{3.17}$$

As expected, the asymptotic UJM has no long-range correlations between opposite rapidity hemispheres. Furthermore, for any scaling matrix element, the inclusive rapidity distribution is essentially given by

$$\frac{1}{\sigma}\frac{d\sigma}{dy} \cong \tilde{\kappa}\left(1 - \frac{2\langle m_\perp\rangle}{\sqrt{s}} \cosh y\right)^{\tilde{\kappa}-1} \tag{3.18}$$

resulting in a rapidity plateau of length $O[\ln(S)]$ with a shoulder width of roughly 2-3 units in y (Fig.3.1).

One further feature of inclusive spectra in the UJM is worth noting. Equation (3.7) suggests a factorization of $E\, d^3\sigma/dp^3$ in terms of y and $p_\perp$. Due to the factor E in the invariant momentum space volume, this does not hold if one replaces y by other longitudinal variables as x or $m_{\shortparallel} = (m^2+p_{\shortparallel}^2)^{1/2}$. This is clearly demonstrated in Fig.3.2, where $\langle p_\perp\rangle$ is plotted vs $m_{\shortparallel}$.

This "seagull effect" is observed in e^+e^- annihilations /24/ as well as in hadron-hadron interactions /40/ and is often interpreted as a direct evidence for a parton structure of the produced particles /56-58/. One should, however, be aware that part of the effect may be simple jet kinematics.

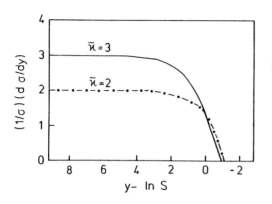

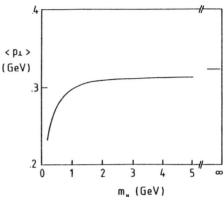

Fig.3.1. Asymptotic rapidity distributions in the UJM ($<m_\perp> = 0.35$)

Fig.3.2. Mean transverse momentum of secondaries as a function of the longitudinal mass for $\lambda = 6.2$ GeV^{-1} in the UJM

3.2 The Approach to Scaling

At typical SPEAR- or DORIS-energies, where the bulk of e^+e^- data was produced, masses and transverse momenta are not a priori negligible. The approach to scaling /54/ will therefore be discussed in more detail.

Obviously, the scaling limit of $d\sigma/dx_R$ is reached latest for x_R close to 0 and 1. At $x_R \lesssim <m_\perp>/\sqrt{s}$, scaling is violated because of threshold effects; in the vicinity of $x_R \cong 1$ (for $\ln^{-1}[(S/4m^2)(1-x_R)] \gtrsim 1$) the unobserved system has a low missing mass M_L (3.10) and hence $\Omega(\hat{\underline{e}},Q-p)$ (3.5) is not asymptotic. Qualitatively, scaling in $d\sigma/dx_R$ will be approached from below at low x_R and from above at high x_R. To study the approach to scaling quantitatively, it is more convenient to calculate $\Omega(\hat{\underline{e}},Q)$ by numerical integration /54,59,60/. The results for $(1/\sigma)(x_R d\sigma/dx_R)$ are compared in Fig.3.3 for $\sqrt{s} = 3.0$, 3.8, 4.9 and 20 GeV with the asymptotic limit, using the "standard" choice of parameters $\lambda \cong 6$ GeV^{-1}, or $<p_\perp> \cong 0.33$ GeV, and $\tilde{\kappa} = 3$ /54/. Although there seems to be an early scaling already at $\sqrt{s} = 3-5$ GeV for $x_R > 0.2$, the calculation shows that one is still far from the scaling limit, which is reached within 0 (10%) at $\sqrt{s} = 20$ GeV for $0.05 < x_R < 0.8$.

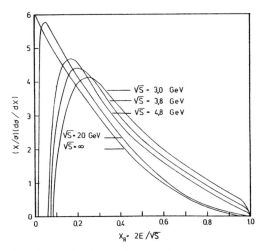
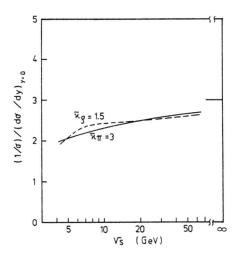

Fig.3.3. The inclusive single-particle spectra in the UJM for $\lambda = 6.2$ GeV^{-1}, $\tilde{\kappa} = 3$ /54/

Fig.3.4. Normalized particle densities at $y = 0$ in the UJM. (———) direct production of $\pi, \tilde{\kappa}_\pi = 3.$ (---) production via ρ decay, $\tilde{\kappa}_\rho = 1.5$ /61/

Figure 3.4 shows the height of the rapidity plateau at $y = 0$ as a function of $\sqrt{s}$ /61/. At energies of $\sqrt{s} = 3$-10 GeV, the particle density is ~30% below the asymptotic value of (3.8); even at the highest energies available at present accelerators (ISR) the UJM predicts the plateau height not to be constant.

Another nonasymptotic effect, namely the influence of the production of heavy particles like K or p, becomes evident if one tries to compare the UJM model with data from e^+e^- storage rings.

Of course, the UJM is not able to give a realistic description of the charge, spin or isospin structure of e^+e^- final states. However, with appropriate constants κ and λ it should describe the gross features of inclusive spectra averaged over particle types, quantum numbers, etc. In Fig.3.5a, the UJM prediction for $\tilde{\kappa} = 3$ and $\lambda = 6.2$ GeV^{-1} is shown together with inclusive $\pi^\pm$ spectra $(s/\beta)(d\sigma/dx_R)$ in e^+e^- annihilations at $\sqrt{s} = 4.46$-4.90 GeV /36,62/. The UJM values have been evaluated for $\sqrt{s} = 4.8$ GeV using numerical techniques. There is a strong disagreement between the predicted and the measured x_R slope; furthermore, as we have seen in Chap.2, the high-x_R pion cross sections scale over the whole DORIS/SPEAR and PETRA energy range, in contrast to the model.

In principle, the first point could be cured simply by increasing $\tilde{\kappa}$ (3.8). This, however, leads to multiplicities incompatible with experiments (3.9).

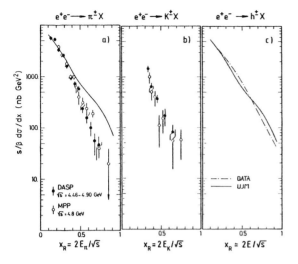

Fig.3.5

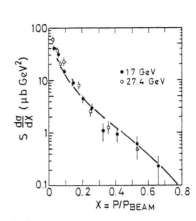

Fig.3.6

Fig.3.5a-c. Comparison of UJM predictions (———) with data from e^+e^- storage rings /36,62/. (a) Inclusive $\pi^{\pm}$ spectrum; (b) inclusive $K^{\pm}$ spectrum; (c) UJM compared to the sum (-·-·-) of the $\pi^{\pm}$ (a), $K^{\pm}$ (b) and $p,\bar{p}$ spectra, after subtraction of charm contributions. The normalization is arbitrary

Fig.3.6. Comparison of the asymptotic limit of the UJM with data from e^+e^- annihilations at $\sqrt{s} \geq 17$ /24/. The UJM parameters are $\tilde{\kappa} = 3$, $\lambda = 6.2$ GeV^{-1}, as in Fig.3.5. The normalization is arbitrary

There is a more natural explanation possible. First, at $\sqrt{s}$ = 4.8 GeV charm production contributes a sizeable fraction to the pion cross section below $x \simeq 0.2$, /36/ which has to be subtracted. Second, in the UJM all particles were assumed to be pions, whereas in practice, also kaons and protons are produced. At moderate $\sqrt{s}$, the rest masses of these particles eat up a good fraction of the available energy, resulting in a steeper decrease of pion spectra at high x. A better (however not fully legitimate) way of comparison is to add up the pion (Fig.3.5a), kaon (Fig.3.5b), and proton spectra (Fig.3.5c). If one further notes that systematic uncertainties between different experiments /35,36,62/ are of the order 2 at high x_R, the agreement is quite impressive.

The early scaling of the experimental pion x_R spectra can thus be interpreted as a cancellation of two nonscaling effects: the production of "heavy" particles and the nonasymptotic $\Omega(\hat{e},Q)$.

At higher energies, $\sqrt{s} \geq 17$ GeV, the UJM is in rather good agreement with data from e^+e^- annihilations (Fig.3.6).

3.3 The UJM and Short Range Correlations

The UJM discussed in the previous chapters describes the production of identical particles without any short range correlation. There is, however, strong evidence from experiments studying particle production in hadronic interactions /63/ that the factorization property of the transition matrix

$$f_N(p_1, p_2, \ldots, p_N, S) = \prod_{i=1}^{N} f(p_i, S) \tag{3.19}$$

holds only in the limit of large invariant masses

$$(p_i + p_j)^2 \gg 1 \text{ GeV}^2$$

whereas low-mass particle systems show strong correlations. Part of these correlations are explained by the cluster model /64/, which states that the primarily produced entities are not the final state particles, but instead are excited states which in turn decay isotropically into stable particles.

The most natural way to enclose short range correlations in the UJM is to assume that the intermediate states, called clusters, are emitted independently. In a second step, charge, energy and momentum conservation in the cluster decay provide low-mass correlations.

Experiments suggest that the bulk of pions is produced via intermediate vector mesons /65/. As an example, let us take the ρ meson as a representative "cluster". Figure 3.7 shows UJM results for emission of ρ mesons decaying into two pions compared to direct production in the scaling limit $S \to \infty$. The ρ meson density is chosen such as to give the same plateau height as for direct emission. Figure 3.7 demonstrates that the inclusion of cluster decays does not alter very much the basic UJM predictions, as far as inclusive spectra are concerned. However, the introduction of the new (cluster mass) scale further delays the approach to the scaling limit /61,66/.

A second source of correlations is related to the quantum nature of secondaries /67/. As far as the emission of identical particles is concerned, the invariant momentum space representation of the transition matrix (3.1) is not fully adequate /68/, since these final state particles obey Boltzmann statistics. In reality, pion production for example is governed by Bose-Einstein (BE) statistics, leading to an "attraction" of like-sign pions, since two pions "like" to be in the same quantum state. The range in momentum space of this type of dynamical correlation can be calculated by noting that two particles are in the same quantum state if

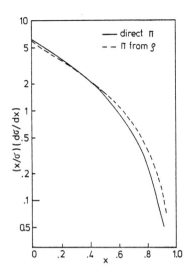

Fig.3.7. x distributions in the UJM for direct, and resonance production of pions. The matrix elements are chosen to give identical heights of the rapidity plateau

$$|(\underline{p}_1-\underline{p}_2)|R \leq \hbar = 1, \text{ or } |\underline{p}_1-\underline{p}_2| \lesssim 200 \text{ MeV}$$

R is the uncertainty of the point of emission, or in other words, the radius of the particle source, which is expected to be of the order 1 fm $\cong$ (200 MeV)$^{-1}$.

An alternative derivation of this result has been given by COCCONI /69/ and by KOPYLEV and PODGORETZKI /70/ using higher-order interference effects in analogy to optics.

Such a correlation modifies the usual phase space weights in such a way as to favor events which have high particle numbers in identical states. At large rapidities or transverse momenta, this tendency is balanced by the finite amount of energy available; therefore, the enhancement of inclusive densities is strongest at low $p_\perp$ and y /71/.

Models of this type have been used to explain the slight rise in $d\sigma/dy$ at y = 0 in the ISR energy range /72/. Nevertheless, one should keep in mind that conventional phase space models also predict a nonasymptotic behavior at those energies.

Anyhow, these results should be taken qualitatively rather than quantitatively. In reality, a lot of physically different particles are produced, which of course show no BE attraction. Furthermore, the model assumes that all particles are produced in the same volume R^4 in space-time because otherwise the particles could be labeled by their production coordinates (within the limits imposed by the uncertainty relation) and hence were not identical. We shall see later (Chap.4,5), that in quark jets, for example, a BE interference is nearly impossible because of this last argument.

3.4 Summary

The uncorrelated jet model for independent emission of particles enables us to study the main implications of four-momentum conservation in jet production. The explicit addition of short-range correlations is possible via resonance or cluster production. Irrespective of details of the model, and nearly independent of the matrix elements involved, the asymptotic scaling limit is fairly well described by

$$\frac{x_R}{\sigma} \frac{d\sigma}{dx_R} = 2\tilde{\kappa}(1-x_R)^{\tilde{\kappa}-1}$$

The approach to scaling is rather slow, even at PETRA or ISR energies the scaling limit is not fully reached.

4. Jets and Parton Models

In the last decade, the quark-parton model /1-4,73/ has proven to be one of the most useful and intuitive concepts in high-energy physics. In the quark-parton model, hadrons are composed of partons. The binding forces of partons are governed by time scales of the order $\tau \cong 1/m$ with m being a typical mass of a few hundred MeV. In a moving hadron, the time scale is dilated by the Lorentz factor γ. Then we have two main classes of hadron interactions in the quark-parton model. If the interaction time τ_I of hadrons is large compared to $\gamma\tau$, all partons participate coherently in the interaction. If τ_I is small compared to $\gamma\tau$, individual partons may interact. In the asymptotic limit $\tau_I \to 0$ the hadron seems to be composed of free, on-shell quarks. The consistency of the model requires the partons to carry fractional charges, and a new quantum number, color /17,74,75/. Since no fractional-charged objects have been observed up to now, there must be a nonasymptotic mechanism confining partons to integer-charged states with the observed hadron quantum numbers. These states turn out to be color singlets. Meanwhile, it became customary to turn these arguments upside down and to consider color confinement as the primary mechanism, and the nonexistence of fractionally charged particles as a secondary consequence.

Consider now a process where a large four momentum is transferred to one parton out of a color singlet system. Obviously, due to the large invariant mass of the parton final state, color cannot be confined to a single stable hadron anymore. One of the basic assumptions of the quark-parton model is that in such a case the confinement of color leads to a bunch of particles moving essentially along the direction of the scattered parton and that this process happens with the probability 1.

In this chapter, we shall discuss the dynamics of jet formation in more detail, starting with the simplest type of jets: quark jets produced in e^+e^- annihilations.

4.1 Jets from Quark Confinement

Confinement is most easily visualized in terms of a QCD-supported bag model /76/. QCD - Quantum Chromodynamics - /17/ is the most promising candidate for a theory of strong interactions. It is a nonabelian gauge theory in which the interactions between colored quarks are mediated by 8 colored gauge bosons, the gluons. The 3 color charges are generators of an unbroken $SU(3)_c$ symmetry. QCD is an asymptotically free theory, its dimensionless coupling constant α_s/π tends to zero as the energy increases, allowing a perturbative expansion. At low energies, $Q^2 \simeq 0$ (1 GeV^2), the coupling constant diverges. This low energy behavior of QCD is hoped (but not yet proven) to explain confinement: recent field theoretical investigations indicate that the QCD ground state is a nontrivial two phase vacuum /77/. In the normal phase, outside hadrons, color fields cannot propagate. This effect is known as "chromodynamic Meissner effect" in analogy to the propagation of magnetic fields in a superconductor. There is further an abnormal phase which may contain quarks and gluons as quasi-free particles. Since surface and volume energy is required to create the abnormal phase, the abnormal regions form little bubbles, or "bags" within the normal phase. The dynamics of quark and gluon fields inside a bag is governed locally by the field equations of QCD. It can be shown that such a bag, embedded in the physical vacuum, is stable in its time evolution; the volume energy B (or vacuum pressure) and surface energy S are balanced by the pressure exerted by the gluon fields which are reflected at the phase boundary /76/. B and S can be arranged to give typical bag sizes of 1 cubic fermi. Such a bag containing a quark and a antiquark or three quarks can be identified as a meson or as a baryon, respectively. Actually the spectrum of bag excitation modes reproduces the observed

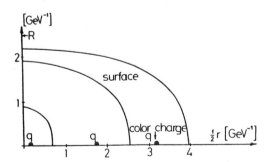

Fig.4.1. The shape of one quadrant of a typical bag /76/ for different quark-antiquark separations. The arrow indicates the radius of an ideal cylindrical vortex tube

hadronic mass spectrum fairly well /76/. What happens, if a large amount of momentum is transferred to a quark inside a bag? As the quark moves away, the bag changes its shape from a sphere to a cylinder (Fig.4.1), and the kinetic energy of the quark is converted into surface and volume energy of the bag. When the bag reaches a critical length, it is energetically more favorable to create a new quark-antiquark pair in between the initial quarks. These new quarks screen the inititial color fields and the bag breaks up into two new, spherical bags, each containing one initial and one new quark. One of these bags is a slowly moving ground state bag, while the other, which contains the fast quark, starts to become cylindrical, repeating the process until the whole initial energy is converted into a series of bags moving along the direction of the initial momentum transfer. The transverse momentum with respect to the jet axis of these bags is determined by the transverse bag size.

$$<p_\perp> \cong O(1 \text{ fermi}^{-1}) \cong O(200 \text{ MeV})$$

4.2 Space-Time Development of Quark Jets

The confinement mechanism proposed above, via successive polarization and de-excitation of the vacuum has been investigated already in 1962 by SCHWINGER /78/. He showed that in certain gauge theories of charged fermion fields, the only asymptotically stable particles are massive neutral bosons. In such theories, electric charge is a confined quantum number, in analogy to the (yet unproven) color confinement in QCD. This phenomenon occurs in two-dimensional (space-time) QED. It may also happen in four-dimensional gauge theories if the coupling constant exceeds a certain critical value /78/.

Examples of field theories exhibiting the Schwinger phenomenon have been discussed by CASHER et al. /79/. In the following, we shall use two-dimensional QED /78,79/ to give a quantitative description of the process of jet formation. Of course, we are aware of the two essential limitations of this picture, the use of only one space dimension and the absence of photon self couplings in QED. On the other hand, if we keep the attitude of the QCD bag model discussed in the previous section, the nonabelian nature of QCD is mainly responsible for the chromodynamic Meissner effect which in turn reduces the problem to a one-dimensional one, the dynamics of a color string.

We shall proceed as follows. First, jet development is discussed using the formalism of two-dimensional QED. Next, we try to transform these results into an intuitive physical picture. Finally, a simple graph technique is presented, which further illustrates the mechanisms of jet development and enables quantitative predictions.

1+1 dimensional (space-time) QED is defined by the Lagrangian density

$$L = \bar{\psi}i\gamma^\mu \partial_\mu \psi - \frac{1}{4}F_{\mu\nu}F^{\mu\nu} - g\bar{\psi}\gamma^\mu\psi A_\mu \tag{4.1}$$

This model is exactly solvable. It has already been mentioned that in this model there are no asymptotic fermion states; the only stable particle is a boson of mass m

$$m = g/\sqrt{\pi} \tag{4.2}$$

The equation of motion in the presence of external current sources is given by

$$(\Box + m^2)j^\mu(x) = -m^2 j^\mu_{ext}(x) \tag{4.3}$$

Let us consider the simplest case, the production of a q-q̄ pair by a virtual photon. The external currents in the photon rest system (which correspond to color currents in QCD) are defined by

$$\begin{aligned} j_{0ext} &= g\delta(z-t) + g\delta(z+t) \\ j_{1ext} &= -g\delta(z-t) + g\delta(z+t) \end{aligned} \tag{4.4}$$

(In our notation, x is a two-vector with the components $x_0 = z$, $x_1 = t$).

It is now convenient to express the current in terms of a dipole density ϕ

$$j^\mu = \epsilon^{\mu\nu}\partial_\nu \phi \tag{4.5}$$

in analogy to the well known equations

$$\rho = \frac{\partial \phi}{\partial z}, \quad j = -\frac{\partial \phi}{\partial t} \tag{4.6}$$

The external dipole density is then

$$\phi_{ext} = -g\theta(t+z)\theta(t-z) \tag{4.7}$$

and the induced polarization density satisfies

$$(\square + m^2)\phi = gm^2\theta(t+z)\theta(t-z) \tag{4.8}$$

The resulting dipole density is constant on hyperbolas in space-time, vanishing near the light cone and approaching a constant for distances $|x| > m^{-1}$ from the origin. As the hyperbolas approach the light cone, the regions containing the polarization charge are confined to a length of the order $(tm^2)^{-1}$. The polarization charge combines with the outgoing charge to form a neutral boson as soon as their distance becomes of the order m^{-1} in their common rest frame. In the cms frame this happens after a time t_c

$$t_c \cong \sqrt{Q^2}/m^2 \tag{4.9}$$

The existence of the induced dipole density is equivalent to the creation of charge-anticharge bound states. The momentum distribution of these bosons can be calculated from the field ϕ. One obtains finally

$$\frac{1}{\sigma} E \frac{d\sigma}{dp} = \frac{1}{\sigma} \frac{d\sigma}{dy} = \frac{1}{(2\pi)^2} |\int d^2x \exp(ipx) g\phi_{ext}|^2 = \frac{g^2}{\pi m} = 1 \tag{4.10}$$

The time evolution of the main quantities ϕ, $d\sigma/dy$ and ρ is summarized in Fig.4.2.

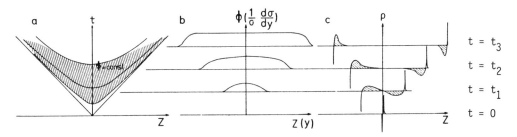

Fig.4.2a-c. Development of charge confinement in the Schwinger model. (a) Lines of constant polarization density in space-time. Particles are produced in the shaded regions. (b) Spatial distribution of the polarization density at different times $(0 > t_1 > t_2 > t_3)$. This is equivalent to the rapidity distribution of the emitted particles. (c) Density of charge at different times. (The initial quarks are represented by delta functions.)

How can we interpret these results? Particle production happens in the cloud of polarization charge, which is induced by the primary charge, i.e., color sources. Particle creation starts at low rapidities. As time goes on, the initial quark feeds energy into accelerating the polarization cloud until the cloud overtakes the leading quark and neutralizes its charge. One should note that the motion of the polarization cloud proceeds through the creation of new quark pairs in front of the cloud and through the recombination of quark pairs into neutral bosons at their end; it need not be a unique quark in the cloud which is accelerated.

The process of charge confinement (or "quark fragmentation") turns out to be characterized by a rather long time scale (4.9). Furthermore, the cascade extends over large distances in space, as compared to typical particle sizes (Fig.4.2); the extension of the cascade is proportional to the quark energy. The maximum distance between the quark charge and the screening polarization charge, however, stays finite at all times (Fig.4.2c).

There is one difference between QCD and QED model worth noting. In QED, the charge of the leading quark is constant during the whole process. In the nonabelian QCD, the colors of quark and polarization cloud change due to gluon exchange, the quark color being opposite to the cloud color in the average. The whole system — quark, cloud, and gluons between them — is a color singlet.

The proportionality between quark momentum and confinement time is important since it ensures that at high energies the outgoing fermions remain free sufficiently long to justify calculations based on the naive parton model.

This model has first been advocated by BJORKEN /80/. Since the slowest hadrons are produced first in time, it is known as an "inside-outside" cascade.

Of course it is questionable to what extent details of the model should be taken seriously and whether it can be generalized to four dimensions. It has been shown that if the Schwinger phenomenon occurs in the four-dimensional field theory, the model gives a rapidity plateau and an appropriate $p_\perp$ cutoff /79/. The problem is, however, to prove that the Schwinger mechanism works, e.g., in QCD. Nevertheless, CASHER et al. /79/ succeeded to prove that the inside-outside evolution of the cascade is a feature common to all fragmentation models describing color confinement. Their argument is quite simple. Imagine a model where the cascades start at the two initial quarks (e.g., via a multiperipheral chain of low momentum transfers). To neutralize their residual charges, the two cascades have to join somewhere in the middle. Since the fragmentation chain represents a random walk, in transverse momentum space and due to the long time scales involved, the probability that the ends of the two cascades overlap in space-time goes to zero as Q^2 goes to infinity. Quarks of energies >> 1 GeV would not be confined in such a model.

Recently, ANDERSSON et al. have presented a semiclassical model for quark-jet fragmentation /81/, which incorporates all features of the Schwinger model and gives a reasonable description of the jets observed in deep inelastic lepton scattering

/83/. In their model, quarks are treated as classical particles, which move on well-defined trajectories in two-dimensional space-time. It is easily seen that the classical treatment is justified as far as longitudinal motions along the jet axis are concerned /84,85/. The relative momentum smearing due to the uncertainty relation

$$dy \cong \frac{dp}{p} \cong \frac{\hbar}{Rp} \; ; \quad R \cong \text{quark bag radius} \tag{4.11}$$

is small if fast quarks are considered, e.g., rapidity and position in space of a quark are well defined simultaneously. Furthermore, one may speculate that interference effects are suppressed since the various particle production points will turn out to be well separated in space-time. The use of only one space dimension is based on the assumption of a "QCD-Meissner effect"; more precisely it means that no transverse degrees of freedom of the elongated quark bags ("flux tubes") are excited /86/. Since there is only one space dimension, the force acting between two oppositely charged (or colored) partons is independent of their distance and is attractive.

In the zero-mass limit, the world lines of the two valence quarks inside a meson are described by Fig.4.3a: The quarks start to separate, losing momentum at a rate

$$dp/dt = \pm g^2/4\pi \tag{4.12}$$

After a certain time, the motion is reversed, the quarks move together, start to separate again and so on. Using the uncertainty relation to relate the mean quark separation and momentum, one obtains for the mass of the meson ground state /81/

$$m \sim g/\sqrt{\pi}$$

The constant g is related to the Regge slope α' via $g^2 = 2/\alpha'$ /85/, or can be taken from linear potential charmonium models /82/, yielding meson masses of the order of a few hundred MeV. In a moving meson, the period of oscillation becomes time-dilated (Fig.4.3b).

In this model, jet formation is visualized in diagrams like Fig.4.4. Two initial quarks move in opposite directions. As soon as a certain energy is stored in the color field (shaded area), the field breaks up somewhere and produces a quark pair.

The break up can be visualized as a classical tunneling effect /85/: somewhere in the color flux tube a virtual quark-antiquark pair is generated by a fluctuation.

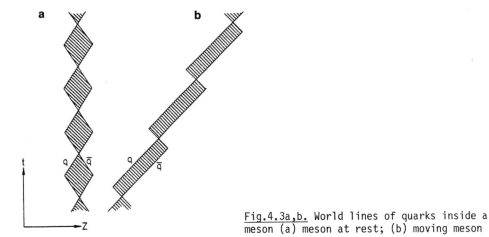

Fig.4.3a,b. World lines of quarks inside a meson (a) meson at rest; (b) moving meson

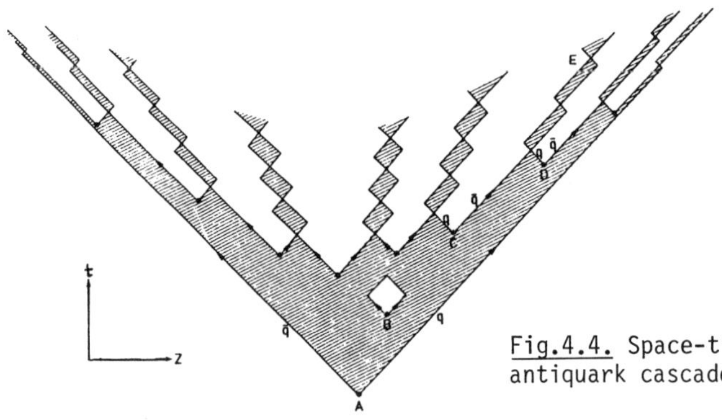

Fig.4.4. Space-time development of a quark-antiquark cascade

The pair moves in a potential

$$E = -2E_T + \frac{g^2}{4\pi} d \qquad (4.13)$$

where E_T is the transverse energy $E_T = \sqrt{(p_\perp^2+m^2)}$ of each of the new quarks, and d is the separation between quark and antiquark. This second term arises since the new quarks screen the external color field in between them. The pair is assumed to be created with zero net momentum since a system with negative total energy cannot propagate with real momentum components /85/.

If the quarks succeed to tunnel from $d = 0$ to $d = 8\pi E_T/g^2$, their energy becomes positive and they materialize as two on-shell quarks. In the WKB approximation the probability per unit time per unit volume to produce on-shell pairs with transverse momentum p_T is given by /85,86/

$$\frac{d\sigma}{dp_\perp^2} \sim \frac{g^2}{8\pi^4} \exp\left(-\frac{4\pi^2 E_T^2}{g^2}\right) \tag{4.14}$$

As soon as kinematically allowed, new pairs are produced in the region between the initial quark and a quark of the first pair, etc. Quarks and antiquarks produced at different points C and D in space-time may join to form a meson E. This is possible, if at the time where quark and antiquark meet in space, the $q\bar{q}$ invariant mass equals the meson mass, or equivalently

$$(Z_C - Z_D)^2 - (t_C - t_D)^2 = m^2 \left(\frac{4\pi}{g^2}\right)^2 \tag{4.15}$$

Equation 4.15 determines the minimum space-time distance of the production points of quarks appearing in final state mesons.

As a consequence of (4.15), particles are emitted from a hyperbola-like space-time region as in the Schwinger model. Two remarks for the sake of completeness will be made:

- In principle it is possible that a quark pair has a large invariant mass $m \gg 1$ GeV. This would split the event into four separate jets. This processes will be rare; if the coupling constant is large enough, quark pairs are created and recombined as soon as possible.
- Figure 4.4, as well as (4.12-14) are slightly oversimplified. Actually quarks are produced in three colors, only one of which screens the external color field. However, it has been shown that the production of the "screening" color is favored and one may argue that the production of other colors leads to a configuration which finally discharges the color field via baryon production /85/.

On the average, the slowest particles are produced first. There is, however, no strict mapping between the rapidity and the time of emission of a particle. Only the average production time increases with rapidity.

Such statements on rapidities and times are not Lorentz invariant. Usually they will be invalidated by a change of the reference frame. ANDERSSON et al. /81/ have investigated this point in detail. Their conclusion is that these statements are true in any reference frame, proving the self consistency of the model.

Define now $D_a^{b\bar{c}}(z)$ as the density of mesons with the valence quark flavor $b\bar{c}$ in the fragmentation region of quark a. z is the fractional momentum of the meson. Masses and transverse momenta are neglected for the moment.

Averaging over all possible diagrams like Fig.4.4, the model yields a set of coupled integral equations for the D's

$$D_a^{bc}(z) = \delta_{ab}f_c + \sum_d \int_z^1 \frac{dz'}{z'} f_d D_d^{bc}(\frac{z}{z'}) \tag{4.16}$$

f_c denotes the probability for a color field to create a pair of quark flavors $c\bar{c}$. The break up is assumed to happen with equal probability anywhere between the quarks generating the field, otherwise f depends on z or z'. Color and spin degrees of freedom are implicitly summed over, with the restriction that $b\bar{c}$ is a color singlet. Simplified to the case of one flavor, (4.16) reads

$$D(z) = 1 + \int_z^1 \frac{dz'}{z'} D(\frac{z}{z'})$$

Equation 4.13 is easily solved to give

$$D(z) = \frac{1}{z} \text{ or } \frac{1}{\sigma}\frac{d\sigma}{dy} = 1 \tag{4.17}$$

Equation 4.17 holds only for production of quarks with negligible constituent masses and transverse momenta. The inclusion of these effects changes the behavior of D(z) from

$$D(z) \to 1 \text{ to } D(z) \to (1-z)$$
$$z \to 1 \quad\quad\quad z \to 1$$

The reason is easily seen. For nonzero masses, quark and antiquark of a pair materialize at a distance d ~ E_T (4.13). However, a break up cannot happen with equal probability anywhere along a flux tube. A minimum average distance d/2 between the center of a pair and the end of a flux tube is required, yielding a reduced number of large-momentum secondaries.

One must also keep in mind that the mesons described by (4.16,17) are not the observed stable hadrons. According to the spin weights, 3/4 of them are vector mesons decaying into a number of pions. This results in a smearing of the rapidity distribution, it further increases the height of the plateau.

Adjusting the f's (4.16) to give a suppression of heavy-quark production, as predicted on the basis of (4.14) /178/

$$f_{up} = f_{down} = 0.4; \; f_{strange} = 0.2; \; f_{charm} = f_{bottom} = \ldots = 0 \tag{4.18}$$

(4.16) can be solved numerically, taking into account resonance decays, etc.

The results are in rough agreement with the data (see, e.g., Fig.4.5), although recent experiments indicate that the amount of SU(3) breaking (4.18) is still somewhat underestimated /162,179/.

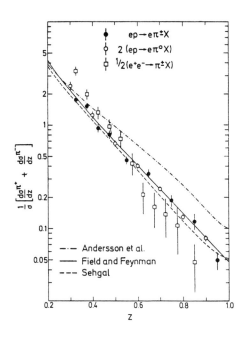

Fig.4.5. Predictions of various jet models /86,87,90/ compared to quark fragmentation functions measured in deep inelastic lepton-nucleon reactions /91/

4.3 An Algorithm for Simulation of Quark Jets

To provide a standard jet model representing the state of the art, FEYNMAN and FIELD /87/ have proposed a similar jet model based on parton phenomenology. The model uses a recursive generation principle originally introduced by KRZYWICKI and PETERSSON /88/ and by FINKELSTEIN and PECCEI /89/. The jet model is intended to serve as a reference for design and comparison of experiments. It generates exclusive multi-particle final states including resonance production, transverse momenta and finite-energy effects. Furthermore, it allows detailed predictions on two- and more-particle correlations and on their quantum-number dependence. The main difference to the model discussed above is that the Feynman-Field (FF) model involves four free parameters adjusted to fit the data: 1) one arbitrary function ultimately determines the momentum distribution of the hadrons, 2) the degree that SU(3) is broken in the formation of new quark-antiquark pairs, 3) the spin of the primary mesons and 4) the mean transverse momentum given to this mesons. In the original version of the model /87/, the production of baryons and of heavy-quark flavors, like charm or beauty is neglected. The ansatz is based on the idea that a quark of type "a" coming out at some momentum p_0 creates a color field in which new quark pairs are produced. Quark "a" then combines with an antiquark, say "$\bar{b}$", from the new pair "$b\bar{b}$" to form a meson "$a\bar{b}$" leaving the remaining quark "b" to combine with a next quark "$\bar{c}$" and so on. The primary meson "$a\bar{b}$" may be directly observed as a stable meson, or it may be an unstable resonance which subsequently decays. A hierarchy of primary mesons is formed in which "a$\bar{b}$" is first in rank, "b$\bar{c}$" is second in rank, "c$\bar{d}$" is third in rank, etc, as shown in Fig.4.6 (here and in all further diagrams we have adopted the convention to plot antiquarks as quarks moving backwards along their world lines). The chain decay ansatz assumes that if the primary meson of rank 1 carries away a momentum ζ_1, the further cascade starts with a quark "b" with the momentum $p_1 = p_0 - \zeta_1$ and the remaining hadrons are distributed in the same way as if they came from a jet originated by a quark of type "b" with momentum p_1. It is assumed that the momenta are high enough, so that all distributions scale.

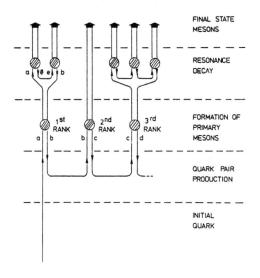

Fig.4.6. Production of a cascade via successive q$\bar{q}$ creation and recombination

Given these assumptions and defining f(z) by f(z)dz = probability that the rank 1 primary meson carries a fraction z of the initial quarks momentum, with $\int_0^1 f(z)dz = 1$, the longitudinal structure of the quark jet is uniquely determined. For the single particle density of primary mesons (regardless of its rank)

$$D(z) = \frac{1}{\sigma}\frac{d\sigma}{dz}$$

with

$$z = \frac{p_{meson}}{p_0}$$

we have the integral equation

$$D(z) = f(z) + \int_z^1 \frac{dz'}{z'} f(1-z') D(\frac{z}{z'}) \qquad (4.19)$$

(assuming that $zp_0 \gg m_\perp$). Generalized to many-quark flavors, (4.19) reads

$$D_a^{bc}(z) = \delta_{ab}f_c(z) + \sum_d \int_z^1 \frac{dz'}{z'} f_d(1-z') D_d^{bc}(\frac{z}{z'})$$

[(in the notation of (4.16)].

The first term is the probability for the meson to have rank 1, the second term arises a sum over all higher ranks with the rank 1 meson being at z'. Feynman and Field assume that $f_b(z)$ (the probability that a rank 1 meson "$a\bar{b}$" is formed at z) can be factorized as follows

$$f_b(z) = f_b^0 \cdot f(z) \qquad (4.20)$$

where

$$\int_0^1 f(z)dz = 1 \qquad (4.21)$$

and

$$\sum_b f_b^0 = 1 \qquad (4.22)$$

The breaking of the SU(N) flavor symmetry is put in the f^0's, ideally one has in SU(N)

$$f_b^0 = \frac{1}{N}$$

Guided by experiments, Feynman and Field use

$$f_{up}^0 = f_{down}^0 = 0.4; \quad f_{strange}^0 = 0.2; \quad f_{charm}^0 = f_{bottom}^0 = \ldots = 0 \qquad (4.24)$$

in agreement with (4.18).

In contrast to the model by Andersson et al., the spin structure of the primary mesons is kept as a free parameter. The best fit to the data is quoted for

$$\begin{aligned}
\text{probability for spin 0 mesons} &= \alpha_p = 0.5 \\
\text{probability for spin 1 mesons} &= \alpha_V = 0.5 \\
\text{probability for spin} \geq 2 \text{ mesons} &= \alpha_T = 0
\end{aligned} \qquad (4.25)$$

Of course, the smaller fraction of vector mesons requires a change of f(z) compared to /91/. Feynman and Field use

$$f(z) = (1-a)+3a(1-z)^2 \text{ with}$$
$$a \cong 0.77 \qquad (4.26)$$

As in (4.14), it is assumed that the new quark pairs have no net transverse momentum. Transverse momenta $\underline{p}_\perp$ and $-\underline{p}_\perp$ are assigned to the quark or antiquark of a pair, with a distribution

$$\frac{1}{\sigma}\frac{d\sigma}{dp_\perp^2} \sim \exp[-p_\perp^2/(2p_{\perp 0}^2)] \qquad (4.27)$$

The transverse momentum of a primary meson is given by the vector sum of the transverse momenta of its quarks. To reproduce the observed mean transverse momentum of the final state pions, $p_{\perp 0} \cong 350$ MeV is required. The mean transverse momentum of primary mesons is of the order of 440 MeV.

The recursive scheme of jet generation is now evident:

I) choose the momentum fraction of the rank 1 meson from (4.26)
II) generate a quark pair according to (4.24,27). The rank 1 meson is made of the old quark and the new antiquark
III) decide on spin-parity of the meson (4.25). Use the known pseudoscalar and vector-mixing angles to determine the precise type of the meson. If necessary, let it decay.
IV) If there is still enough energy left over, repeat the procedure starting with the residual quark.

At this stage, a few remarks concerning the philosophy of the model have to be made. Regarding the recursive principle, one is tempted to consider the Feynman-Field jet as an "outside-inside" cascade, where particles of rank 1 are produced first at high y in the overall cms. The authors themselves considered this point as one of the major drawbacks of their model /87/. A comparison of the integral equations (4.19) with the equations describing the inside-outside cascade of ANDERSSON et al. (4.16) however proves that both point of views lead to very similar mathematical structures, the (4.16) being as well compatible with an inside-outside development of the jet. The recursive principle proposed by Feynman and Field thus should not be considered as a physical model, but merely as a mnemonic simplifying the bookkeeping of momenta, quantum numbers, etc.

A second remark is more technical. The FF model describes the fragmentation of a single quark. Consequently, one single antiquark with its charge, color and (small) momentum has to be thrown away at the end of the cascade after n mesons have been produced. In reality, these quantities have to be neutralized by the corresponding ones in the opposite jet. This unbalance of conserved quantities is not a specific failure of the FF technique to generate jets, instead it is a reflection of the fact that it is meaningless to talk about a single jet. The only systems of physical relevance are color singlets, like $q\bar{q}$ or qqq states. A division of such systems into single jets is an approximation possible for fast particles, it is however completely meaningless and arbitrary for slow fragmentation products.

4.4 Properties of Quark Jets

In this section we shall discuss the expected properties of quark jets, and compare them to the experiment. The FF model will be used as the typical parton-jet reference. The jet energy should be chosen such that mass effects are negligible, on the other hand it should be low enough so that the QCD corrections to be discussed in the next section will not dominate. These conditions are fulfilled in the lower PETRA energy range.

Consider the fragmentation of u,d, and s quarks (the decay of charmed quarks will be discussed later). As "stable" final state particles we have

$$u,d,s \rightarrow \pi^+, \pi^0, \pi^-; K^+, K^0, K^-, \overline{K^0}; \gamma \qquad (4.28)$$

In principle, there are 3×8 fragmentation functions (plus the corresponding ones for antiquarks).

$$D_q^h(z) = \frac{1}{\sigma_q} \frac{d\sigma_h}{dz} \qquad (4.29)$$

Parton models allow to reduce the number of independent D's drastically. Essentially, there are two main fragmentation mechanisms, usually known as favored and unfavored fragmentation /90/. A fragmentation is favored if the fragment contains the initial quark. In the language of Feynman and Field it may then be the (fast) meson of rank 1. In the unfavored case, where the meson does not contain the initial quark, it has to be of rank 2 or higher, implying a smaller mean momentum. Since the flavor of the quarks generated in the first break up of the color flux tube should not depend on the nature of the primary quark, all unfavored D's should be equal, except for a constant suppression factor ξ for strange secondaries [according to (4.14)]. Favored fragmentation functions are obtained by adding an universal distribution D_0 of rank 1 mesons to the corresponding unfavored fragmentation function D_{unf}, e.g.,

$$D_u^{\pi^-} = D_{unf}; \quad D_u^{K^-} = \xi D_{unf}$$
$$D_u^{\pi^+} = D_{unf} + D_0; \quad D_u^{K^+} = \xi(D_{unf} + D_0)$$

Of course, these expressions are only approximations because of the presence of a second mechanism for "unfavored" fragmentation: the decay of vector mesons, e.g.,

$$u \rightarrow \rho^0 + \text{anything} \rightarrow \pi^+ \pi^- + \text{anything}$$

which competes with the direct suppressed channel.

All this is more easily demonstrated in Table 4.1, where the fraction of momentum carried by the various decay products is given, and by Fig.4.7, where the favored and unfavored fragmentation functions of a u-quark into π's and K's are shown as predicted by the FF model.

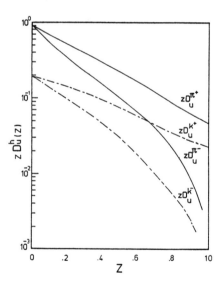

Fig.4.7. u-quark fragmentation function in the FF model

Table 4.1. Fraction of total momentum carried by the mesons resulting from a u-, d- and s-quark /87/

	Particle	u	d	s
	π^+	0.29	0.19	0.19
	π^0	0.26	0.26	0.20
	π^-	0.19	0.29	0.19
	K^+	0.08	0.06	0.06
	K^0	0.06	0.08	0.06
	K^-	0.04	0.04	0.13
	$\overline{K^0}$	0.04	0.04	0.13
	γ	0.04	0.04	0.04
total	π	0.74	0.74	0.58
total	K	0.22	0.22	0.38

Two points are worth noting. In contrast to phase space models (Chap.3), the FF model predicts a nonzero value of $D_q^u(z)$ at $z \to 1$ even at asymptotic energies. On the other hand $D_q^\pi = D_q^{\pi^+} + D_q^{\pi^-}$ is remarkable well fitted by the UJM expression $D(z) = (1-z)^2/z$ in a large z range $0 < z < 0.7$.

In e^+e^- annihilations the fragmentation functions of u-, d- and s-quarks cannot be measured separately. Concerning this topic we shall use data from deep inelastic lepton nucleon interactions for comparison. To reduce contributions from quasi-elastic channels, cuts on the four-momentum transfer Q^2 and on the invariant mass W of the hadronic system have been applied, typically $Q^2 > 2$ GeV^2 and $W > 4$ GeV. Nevertheless, the mean energy in the hadronic center of mass $<W> \cong 6$ to 7 GeV is fairly small, and phase space effects are still important.

As we shall see in more detail in Chap.6, the choice to the scaling variable is somewhat ambigious in lepton-nucleon interactions, especially as far as slow particles are concerned. The most commonly used variable is $z = E/E_{had}$. E is the hadron energy in the laboratory frame, and E_{had} is the total hadronic energy, which is equivalent to the quark energy. For fast hadrons ($z \gtrsim 0.2$), z coincides with $x = 2p_{cms}/W$ or with $x_R = 2E_{cms}/W$. Figure 4.5 shows that the model fits the spectrum of quark fragments rather well as derived from electroproduction /91/. In Fig.4.8 data from neutrino-nucleus and neutrino-proton interactions on the favored and unfavored fragmentation modes of u-quarks are compared to the FF model. Again the agreement is good. Included in Fig.4.8 are predictions of a longitudinal phase space model, which demonstrates the influence of energy-momentum and charge conser-

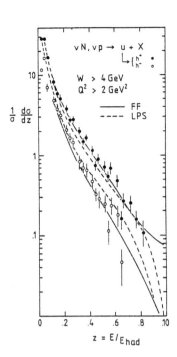

Fig.4.8

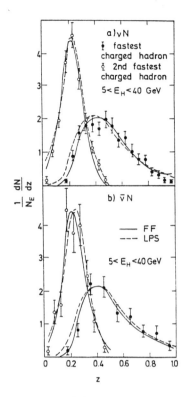

Fig.4.9

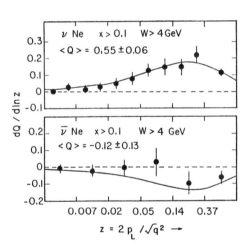

Fig.4.8. u-quark fragmentation functions measured in neutrino nucleus interactions /93/ compared to predictions of the FF model and of a phase space model (LPS)

Fig.4.9a,b. Distribution in z for the fastest and the second fastest charged hadrons, compared with FF and LPS in (a) νN and (b) $\bar{\nu}$N interactions /94/

Fig.4.10. Distribution of the charge $dQ/d\ln(z)$ plotted vs $\ln(z)$ for u- and d-quark fragmentation /95/. The curve is from Feynman and Field

vation. The multiplicity has been constrained to the observed one. Obviously, much of the difference observed between the "favored" and the "unfavored" distribution of positive and negative hadrons is accounted for by simple charge conservation. Even the increase of the difference with z is reproduced. Events with a particle close to $z \cong 1$ have a low missing mass and therefore a low multiplicity; effects due to charge conservation are more important than in average events.

A more differential quantity, the distribution of the fastest and second fastest charged hadron, is shown in Fig.4.9 /94/. The FF model as well as the naive phase space model agree with data. Finally, Fig.4.10 illustrates the net charge density in u- and d-quark jets as a function of the rapidity $y_z = \ln(z)$ /95/. The integrated charge content of the jets is $<Q>_u = 0.55\pm0.06$ and $<Q>_d = -0.12\pm0.13$ for u- and d-quarks, respectively. The FF model has the tendency to reproduce the observed asymmetry between $<Q_u>$ and $<Q_d>$.

To summarize so far, the FF model seems to be in reasonable agreement with data from deep inelastic reactions even in details; however, the comparison with the longitudinal phase space model proves that most of this agreement is required by conservation laws, once the multiplicity distribution is correctly adjusted.

Let us now study in more detail the quantum-number structure of quark jets. We have seen in Fig.4.10 that quark jets tend to retain the charge of the quark at high rapidities. This is the object of a suggestion by FEYNMAN /4/. He conjectured that the total quantum numbers of all the fragments in this region averaged over events should be equal to the (fractional) quantum numbers of the parent quark. A necessary, but not sufficient condition for retention of quantum numbers is that the rank 1 mesons stay at large rapidities in the jet. This is easily demonstrated. Assume we cut the jet somewhere, e.g., at $z_0 = 0.01$ and sum up the quantum numbers of all particles with higher z. In nearly all cases the initial quark is contained among these mesons. Yet, since we start with a quark and count an integer number of mesons, there is always one quark pair whose quark is below z_0 and whose antiquark is above. If z_0 is low enough, this quark pair will stem from the sea. The sum Λ of quantum numbers for $z > z_0$ is then given by

$$<\Lambda> = \Lambda_q + <\Lambda_s> \tag{4.30}$$

Λ_q describes the quantum numbers of the initial quark (or antiquark) and $<\Lambda_s>$ are the mean quantum numbers of an antiquark (or quark) from the sea. Exact quantum-number retention happens only if $<\Lambda_s> = 0$. Let us consider a few examples: a SU(2) sea consisting of $u,\bar{u},d,\bar{d}$; a SU(3) symmetric sea; and the FF quark sea with SU(3) broken due to the higher s-quark mass. We get for $<\Lambda_s>$

Sea	Charge	Strangeness	Baryon number	3rd Component of isospin	
SU(2)	-1/6	0	-1/3	0	(4.31)
SU(3)	0	1/3	-1/3	0	
SU(3)$_{FF}$	-1/15	1/5	-1/3	0	

In none of these models are all quantum numbers retained. For the FF sea, we expect approximate retention of charge and I_z. The mean quantum numbers of their jets are

Initial quark	Mean jet charge	Mean Jet strangeness	Mean I_z	
u	0.60	0.20	0.50	(4.32)
d	-0.40	0.20	0.50	
s	-0.40	0.80	0.00	

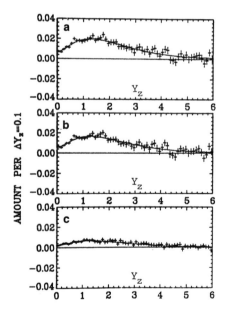

<u>Fig.4.11a-c.</u> Distribution of (a) charge Q, total Q = 0.6, (b) third component of isospin I_3, total I_3 = 0.5, and (c) strangeness S, total S = 0.2, along the $Y_z [Y_z = -\ln(z)]$ axis for a u-quark jet

These numbers are confirmed by the explicit Monte Carlo simulations of jets /87/, the results for quantum-number density in quark jets are shown in Fig.4.11. The most remarkable consequence of these considerations is that even if a jet starts from a nonstrange quark, it finally acquires a net strangeness, which has to be compensated by the opposite antiquark jet. Furthermore, in any jet mechanism leading to a rapidity plateau (that means the decay function f(z) (4.16) is finite for $z \to 0$) only the high-y region of the jet is influenced by the parents quantum numbers, and the plateau essentially stays neutral.

4.5 Summary

Based on the quark-parton model, the space-time evolution of a quark jet can be visualized as an "inside-outside" cascade. Independent of the reference frame chosen, slow fragments are produced earliest. The production of leading fragments, or equivalently the final neutralization of the primary color charge occurs after a time T proportional to the quark momentum Q. A quantitative description of this process can be given using an analogy to two-dimensional QED, or a semiclassical quark model. The recursive scheme for jet generation proposed by Feynman and Field is consistent with these models; it produces the measured inclusive particle distributions in quark jets, and the data on correlations, fairly well. One should note, however, that much of this agreement results simply from constraints due to four-momentum and charge conservation and can be obtained as well in an UJM (Chap. 3), once the mean multiplicity is correctly chosen.

5. Parton Jets and QCD

The quark-parton model was originally developed to provide a useful and simple description of the physics of deep inelastic phenomena /1-4,96,97/. The modern foundation for the parton model is the gauge theory of strong interactions based on the color degree of freedom: QCD. In QCD scaling of the deep inelastic structure functions, which describe the quark distribution within the nucleon, is predicted to be broken by logarithms of $q^2 = -Q^2$. Physically, these scaling violations are due to the emission or absorption of gluons during the hard-scattering process. Thus although the naive parton model, strictly speaking, fails, it is possible to rephrase the parton language by assigning a well specified Q^2 dependence to the parton densities /98/

$$G(x) \rightarrow G(x,Q^2) \tag{5.1}$$

Already through the principle of analytic continuation of the scattering amplitude a Q^2 dependence of the structure functions, describing the distribution of partons in a hadron, also automatically induces a Q^2 dependence of the fragmentation functions, which parameterize the distribution of hadrons in a parton /99,100/.

The dynamical mechanisms leading to this scale breaking are sketched in Fig.5.1. "Before" the quarks emitted, e.g., in e^+e^- annihilations, reach the surface of the confining bag, they radiate gluons which may in turn convert into new quark-antiquark pairs. The bulk of low-momentum gluons is reflected at the bag surface and compensates the "vacuum pressure". Gluons and quarks in the high-momentum tail of the radiation however deform the surface of the bag and create a number of incoherent final state jets. The "parton shower" inside the bag is described by QCD, whereas the final confining stage has to be described by phenomenological models like the FF jet.

Here we shall follow the historical approach and first discuss the mechanisms of scale breaking for the example of structure functions. Finally, we generalize the results to the case of fragmentation functions.

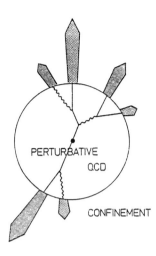

Fig.5.1. The perturbative QCD scale breaking in parton jets through gluon emission and conversion. Full and wavy lines refer to quarks and gluons, respectively

5.1 Scale Breaking in QCD

The formal and theoretically rigorous derivation of scale breaking in QCD is formulated in the language of renormalization group equations for the coefficient functions of local operators in the light-cone expansion for the product of two currents /101/.

A more intuitive model for scaling violations in scale invariant theories has been given by KOGUT and SUSSKIND /44/ and we shall follow their lines of argumentation /44,102,103/. We assume that matter organizes itself into "clusters" /104/. For example, molecules are made of atoms which are made of nuclei which are made of nucleons and so on. Each cluster is characterized by a certain size and time scale. The relation between these scales appears to be accidental. However, as smaller and smaller scales are probed in high-energy physics some kind of regularities may emerge; certain field theories suggest that the connections between adjacent scales may become universal.

Consider time and length scales of ordinary hadrons, 10^{-13} cm, and denote those as $N = 0$ clusters which in turn are composite of $N = 1$ clusters (partons). Renormalization group investigations suggest that the dynamics of clusters of type N which form a cluster of type N-1 can be described by a Hamiltonian H_N without explicitly refering to smaller clusters (N+1, for example). In an infinite momentum frame, H_N depends on the fractional momenta x of clusters of the type N and on their transverse momenta. The intuitive picture suggested by WILSON /105/ and POLYAKOV /106/ was that at large N the coupling constant describing parton-parton interactions becomes constant and thus the ratio of typical scales $R_{N+1}/R_N = \delta$ approaches a constant at large N. In that case, the transformation connecting the Hamiltonians H_{N+1} and H_N is also independent of N.

Imagine now an experiment, which probes hadrons with particles of large four momentum q. Assume the q is spacelike. Then a reference frame can be chosen such that q consists only of a three vector, $q = (0,\underline{q})$. Such a projectile will be able to resolve structures of the order of its wavelength $\lambda = Q^{-1} = (-q^2)^{-1/2}$. As the "target" hadron moves with roughly the speed of light, the time scales probed are of the order Q^{-1} as well. The parton distribution which is relevant for the description of the process is that of clusters with $R_N \cong \lambda$, since clusters of the type N+1 cannot be resolved by the probe, whereas clusters of type N-1 no longer appear pointlike.

This means

$$\lambda \cong R_0 \delta^N \tag{5.2}$$

or, with $\lambda \cong Q^{-1}$: $N \cong -\ln(R_0 Q)/\ln(\delta)$.

To calculate the x distribution of clusters of type N within the original hadron

$$G_N(x) = \frac{1}{\sigma_h} \frac{d\sigma_N}{dx} \tag{5.3}$$

we introduce the function $h_{N+1,N}(z)$ which gives the probability per unit z of finding a cluster of type N+1 and longitudinal momentum zx' in a cluster of type N and longitudinal momentum x'. Invariance under longitudinal boosts requires that h depends only on z and not on the absolute momenta of the clusters. We neglect transverse momenta for the moment being. The distribution of type N+1 clusters is given by the convolution

$$G_{N+1}(x) = \int_x^1 \frac{dx'}{x'} G_N(x') h_{N+1,N}(\frac{x}{x'}) \tag{5.4}$$

We shall now loosen the assumption of the existence of discrete scales; real field theories may not contain such abrupt transitions between scales. A better description is given using a smooth connection between different N and Q. With the replacements

$$\begin{aligned} G_N(x) &\to G(x,t) \\ h_{N+1,N}(z) &\to h(z,t) \\ G_{N+1}(x) - G_N(x) &\to 2\ln(1/\delta) \frac{\partial G(x,t)}{\partial t} \end{aligned} \tag{5.5}$$

and

$$t = \ln(Q^2/Q_0^2) \tag{5.6}$$

we obtain the following equation for $G(x,t)$

$$2\ln(\delta^{-1}) \frac{\partial G(x,t)}{\partial t} = \int_x^1 \frac{dx'}{x'} \left[G(x',t) h(\frac{x}{x'},t) - G(x',t) \delta(1-\frac{x'}{x}) \right] \tag{5.7}$$

or with the redefinition

$$p(z,t) = [h(z,t) - \delta(1-\frac{1}{z})]/2\ln(\delta^{-1}) \tag{5.8}$$

(5.7) reads simply

$$\frac{\partial G(x,t)}{\partial t} = \int_x^1 \frac{dx'}{x'} p(\frac{x}{x'},t) G(x',t) \tag{5.9}$$

Equation (5.9) describes the number density of partons in an object which is probed at a momentum transfer $t = \ln(Q^2/Q_0^2)$. It can be generalized to the case were the probe is only sensitive to the distribution of partons carrying a set of quantum numbers i

$$\frac{\partial G_i(x,t)}{\partial t} = \int_x^1 \frac{dx'}{x'} \sum_j P_{ij}(\frac{x}{x'},t) G_j(x',t) \tag{5.10}$$

P_{ij} is the transition matrix between the various species of partons.

To solve (5.9) it is convenient to rewrite it in terms of moments

$$G(\alpha,t) = \int_0^1 dx\, x^{\alpha-1} G(x,t) \tag{5.11}$$

and

$$p(\alpha,t) = \int_0^1 dx\, x^{\alpha-1} p(x,t) \tag{5.12}$$

Equation (5.9) reads now

$$\frac{\partial G(\alpha,t)}{\partial t} = p(\alpha,t) G(\alpha,t) \tag{5.13}$$

resp. for (5.10)

$$\frac{\partial G_i(\alpha,t)}{\partial t} = \sum_j P_{ij}(\alpha,t) G_j(\alpha,t) \tag{5.14}$$

The structure function $G(x,t)$ can be obtained from its moments via the inverse Mellin transform or by approximative numerical methods /107/.

Let us now try to concatenate this picture and QCD as a description of parton dynamics. We identify a subnuclear "cluster" as a quark-parton, which, when probed at a higher Q^2, looks like one off-shell quark surrounded by a cloud of gluons and quark-antiquark pairs. The coupling between quarks and gluons tends to zero as Q^2 increases; QCD is an aymptotically free theory /74,75/:

$$\alpha_S(Q^2) \cong \frac{1}{b \ln(Q^2/\Lambda^2)} \tag{5.15}$$

with

$$b = \frac{33-2f}{12\pi} \tag{5.16}$$

f is the number of active flavors.

For further calculations we write the transition probability p as a product of the effective coupling constant $\alpha_S(Q^2)/2\pi$ and a term depending only on the momentum ratios

$$p(x,t) = \frac{\alpha_S(t)}{2\pi} p(x) \tag{5.17}$$

Equation (5.13)

$$\frac{\partial G(\alpha,t)}{\partial t} = \frac{\alpha_S(t)}{2\pi} p(\alpha) G(\alpha,t)$$

is now easily solved:

$$G(\alpha,t) = G_0(\alpha) \left[\frac{\alpha(0)}{\alpha(t)}\right]^{p(\alpha)/2\pi b} = G_0(\alpha) \left[\frac{\ln(Q^2/\Lambda^2)}{\ln(Q_0^2/\Lambda^2)}\right]^{p(\alpha)/2\pi b} \tag{5.18}$$

Thus QCD leads to logarithmic violations of scaling. In real QCD, we have not simply one parton-structure function G but instead the structure functions q_i and g of 2f quark flavors ($i = u, \bar{u}, d, \bar{d},...$) and the gluon. Summing over all color states, we get from (5.10):

$$\frac{\partial q_i(x,t)}{\partial t} = \frac{\alpha_S}{2\pi} \int_x^1 \frac{dx'}{x'} \left[\sum_j P_{ij}(\frac{x}{x'}) q_j(x') + P_{ig}(\frac{x}{x'}) g(x') \right]$$

$$\frac{\partial g(x,t)}{\partial t} = \frac{\alpha_S}{2\pi} \int_x^1 \frac{dx'}{x'} \left[\sum_j P_{gj}(\frac{x}{x'}) q_j(x') + P_{gg}(\frac{x}{x'}) g(x') \right]$$

(5.19)

These are the well known ALTARELLI-PARISI master equations /98/. It must be pointed out here that the master equations describe the t dependence of the parton distribution function in momentum space. They do not refer to the time development of parton densities during the scattering process. The p functions can be evaluated using standard graph techniques /98/ (Fig.5.2). In the leading log approximation all terms containing powers in $\alpha_S \ln(Q^2)$ are summed up; the direct terms in P_{ii} and P_{gg} are then given through momentum conservation.

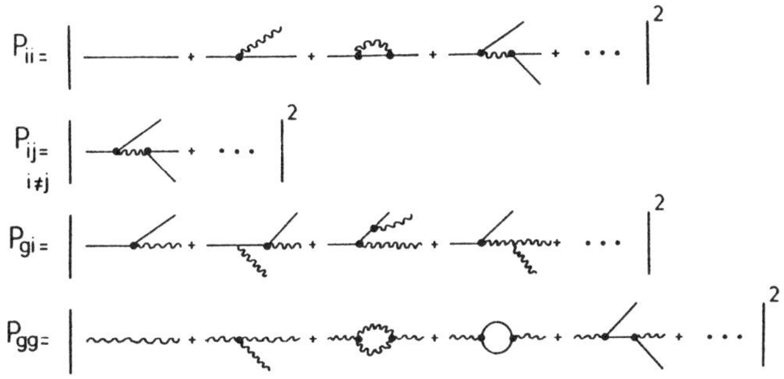

Fig.5.2. First- and second-order contributions to the p functions

$$P_{ij} = \frac{4}{3} \left[\frac{1+x^2}{1-x} \right]_+ \delta_{ij} + 2\delta(1-x)\delta_{ij}$$

$$P_{gi} = \frac{3}{4} \frac{1+(1-x)^2}{x}$$

$$P_{ig} = \frac{1}{2} \left[x^2 + (1-x)^2 \right]$$

(5.20)

$$P_{gg} = 6 \left[\frac{1-x}{x} + (\frac{1}{1-x})_+ + x(1-x) - \frac{1}{12}\delta(x-1) \right] - \frac{1}{3} f \delta(x-1)$$

In terms of moments, (5.19) yields a system of coupled differential equations which can be partially diagonalized by choosing suitable combinations of the parton densities: the flavor singlet and flavor nonsinglet (valence) components $q^S(x,t)$ and $q^{ns}(x,t)$

$$q^{ns}(x,t) = \sum_{u,d,s...} q_i(x,t) - \sum_{\bar{u},\bar{d},\bar{s}...} \bar{q}_i(x,t)$$

$$q^S(x,t) = \sum_{u,d,s...} q_i(x,t) + \sum_{\bar{u},\bar{d},\bar{s}...} \bar{q}_i(x,t)$$

(5.21)

Define now G as the vector of the αth moments of the parton densities

$$G(\alpha,t) = \left[q^{ns}(\alpha,t), q^S(\alpha,t), g(\alpha,t)\right] \qquad (5.22)$$

The master equations read then

$$\frac{\partial}{\partial t} G(\alpha,t) = \frac{\alpha_S}{2\pi} A(\alpha) G(\alpha,t) \qquad (5.23)$$

with the matrix $A(\alpha)$

$$A(\alpha) = \begin{Bmatrix} P_{qq}(\alpha) & 0 & 0 \\ 0 & A_{qq}(\alpha) & 2fA_{qg}(\alpha) \\ 0 & A_{gq}(\alpha) & A_{gg}(\alpha) \end{Bmatrix} \qquad (5.24)$$

The coefficients A_{qq} etc. are combinations of the moments of the p functions. The t dependence of the nonsinglet component is obvious, it is given by (5.18).

Equation (5.23) has been shown to describe the measured nucleon-structure functions, once higher-order corrections and bound-state effects are taken into account /108/.

We are now prepared to discuss the QCD scale breaking of fragmentation functions /109/. Following again the ideas of KOGUT, SUSSKIND /44/ and POLYAKOV /106/, we assume that in a deep inelastic reaction or in e^+e^- annihilation, a parton of type

$$N \cong -\ln(R_0 Q)/\ln(\delta)$$

is emitted. This parton "decays" now in a series of N-1 type partons which decay into N-2 type partons, etc. Consequently, the distance of the partons four momentum from the mass shell increases with N /106/.

In close analogy to (5.4-10) we get

$$G_{N-1}(x) = \int_x^1 \frac{dx'}{x'} G_N(x) k_{N-1,N}(\frac{x}{x'}) \qquad (5.25)$$

or

$$-\frac{\partial G(x,t)}{\partial t} = \int_x^1 \frac{dx'}{x'} r(\frac{x}{x'},t) G(x',t) \qquad (5.26)$$

Since we are still dealing with quark and gluon interactions governed by QCD, the r-functions in (5.26) are identical to the p-functions in (5.19). With the definitions of (5.21,22) we arrive at

$$-\frac{\partial}{\partial t} G(\alpha,t) = \frac{\alpha_s}{2\pi} A(\alpha) G(\alpha,t) \qquad (5.27)$$

For simplicity, we write the moments of the parton-fragmentation functions D as a matrix as well, the element $D_j^i(\alpha,t)$ describing the decay of the jth parton component (nonsinglet, singlet, gluon) into the hadron $i(\pi^+, \pi^0, \pi^- ...)$. The moments of the final state hadron density are then

$$\sigma(\alpha) = D(\alpha,t) G(\alpha,t) \qquad (5.28)$$

There are now two ways to describe the fragmentation of a quark at t_q, as sketched in Fig.5.3. The fragmentation function $D(\alpha,t_q)$ gives a direct mapping of the parton distribution at t_q

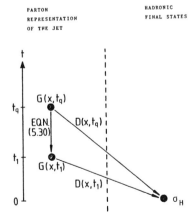

Fig.5.3. The fragmentation of a quark at t_q can be considered as a mapping of the point $G(x,t_q) = [\delta(1-x), \delta(1-x), 0]$ onto a hadronic final state σ_h. This can be done in two steps: $G(x,t_q)$ is first mapped onto $G(x,t_1)$; next $G(x,t_1)$ is mapped onto σ_h

$$G(x,t_q) = [\delta(1-x), \delta(1-x), 0]$$

or

$$G(\alpha,t_q) = (1,1,0) \qquad (5.29)$$

onto the hadronic final state. On the other hand, $G(\alpha,t_q)$ can be described in terms of a superposition of partons at t_1, $0 \leq t_1 \leq t_q$. The relation between $G(\alpha,t_q)$ and $G(\alpha,t_1)$ is given by (5.27). The hadronic final state can now be evaluated by applying $D(\alpha,t_1)$ on $G(\alpha,t_1)$. Of course the final state distribution function has to be independent of t_1. So we obtain

$$\frac{\partial}{\partial t_1}[D(\alpha,t_1)G(\alpha,t_1)] = 0$$

or, with (5.27)

$$\frac{\partial}{\partial t} D(\alpha,t) = \frac{\alpha_s}{2\pi} D(\alpha,t) A(\alpha) \qquad (5.30)$$

To solve (5.30) one needs a boundary condition at a reference Q^2. We could use the FF jet model at $Q^2 \simeq 100$ GeV where QCD effects should be negligible. The explicit calculation of the scale breaking in parton jets is however complicated by the fact that a physical quark or gluon is never a pure flavor nonsinglet. Through the singlet term in the quark density $G(\alpha,t_q)$ quark and gluon terms become mixed up. Thus to find the quark fragmentation function, one has to know as well the gluon fragmentation functions at the reference Q^2. To arrive at a more quantitative understanding of the process, we shall consider the parton densities $G(x,0)$ within a jet initiated by a quark at $Q^2 \gg Q_0^2$. Q_0^2 is chosen as low as possible (of course, the validity of the perturbative QCD expansion has still to be guaranteed). That means, we observe the "parton shower" initiated by a primary quark at a point just before the nonperturbative conversion into hadrons starts. De GRAND /110/ has shown that in the high Q^2 limit the distribution q^{ns} (favored nonsinglet component), q^{unf} (unfavored component, e.g., antiquarks) and g (gluons) from (5.27) may be approximated by

$$q_q^{ns}(x,0) \sim (1-x)^{16\xi/3-1}$$
$$x \to 1, \quad \xi > 0.05$$

$$q_q^{unf}(x,0) \sim (1-x)^{16\xi/3+1} \tag{5.31}$$
$$x \to 1$$

$$g_q(x,0) \sim (1-x)^{16\xi/3}$$
$$x \to 1, \quad \xi > 0.05$$

The subscript q refers to the nature of the parent parton, ξ is defined by

$$\xi = \frac{1}{4\pi b} \ln\left[\frac{\alpha_S(Q_0^2)}{\alpha_S(Q^2)}\right] \tag{5.32}$$

Obviously, as Q^2 goes to infinity, all parton (and hadron) densities concentrate at x = 0. The ratio of the various components stays constant. Compared to the valence quark, gluons and sea quarks are suppressed by factors (1-x) and $(1-x)^2$ respectively. Thus the leading particle effect, which means the dominance at high x of particles containing the original quark is not destroyed by the QCD corrections. The final hadron spectrum is given by the convolution of (5.31) with the fragmentation function $D^h(x,0)$. Therefore, the steepening of the leading hadron distributions occurs at the same rate as in q_q^{ns}. Comparing typical DORIS and PETRA energies, we get

$$\frac{x}{\sigma}\left(\frac{d\sigma}{dx}\right)_{Q^2 \cong 1000 \text{ GeV}^2} \cong (1-x)^{0.5} \frac{x}{\sigma}\left(\frac{d\sigma}{dx}\right)_{Q^2 \cong 10 \text{ GeV}^2} \tag{5.33}$$
$$x \to 1$$

5.2 Preconfinement

If one compares the QCD results on parton fragmentation (5.30) with the results of naive confinement models (Chap.4), a paradox seems to emerge. In QCD, the decay of a parton at Q^2 into many partons of lower Q_0^2 is entirely governed by the color charge of the fragmenting parton. The pictorial flux tube connecting the two color sources plays no role at all, a second source is not even needed! Furthermore, the QCD "final" state at t = 0 contains a number of color triplet and octet sources, which are assumed to decay independently according to fragmentation functions D(x,0), whereas naively one expects a system of color flux tubes joining these sources.

The origin of these problems is the following. The confinement and QCD picture apply to different stages of the process and even to different kinematical regions. The naive confinement picture mainly addresses the question how do nearly on-shell partons transform into hadrons to form a rapidity plateau; whereas QCD describes the evolution of fast partons, which are far off shell within the jet. Nevertheless, the question remains how do the two different approaches, the confinement model at $Q^2 \simeq Q_0^2$, and the QCD picture at high Q^2, join smoothly.

Recently, KONISHI et al. /111/ have proposed a generalization of (5.30) to describe the t-dependence of n-parton cross section in jets. They describe the parton fragmentation as a branching process /112/; an exact derivation of their method has been given by KIRSCHNER /113/ using the leading log approximation. In close analogy to the model of KOGUT and SUSSKIND /44/, it is shown that in the QCD evolution of a jet, partons of size Q_K decay ("branch") into two partons of size Q_{K-1}. The process may be visualized as real decay in spacetime /114/. The lifetime Γ_K associated to the propagator of a parton in the Kth generation is /114/

$$\Gamma_K = x_K Q/Q_K^2 \tag{5.34}$$

in agreement with the naive expectation from the uncertainty relation, $\Gamma_K = \gamma/\Delta E$, with $\gamma = x_K Q/Q_K$ and $\Delta E = Q_K$. From (5.34) we get the total time for jet development

$$\Gamma \simeq \sum_{K=1}^{N} \frac{Q}{Q_K^2} 2^{K-1} \simeq \frac{Q}{Q_0^2} \exp\left[\ln(Q^2/Q_0^2)\right]^{1/2} \tag{5.35}$$

The factor 2^{K-1} in (5.35) arises since the Kth generation contains 2^{K-1} partons. This result is in qualitative agreement with that obtained from the naive fragmentation model of Sect.4.1.

The jet development, by branching of a parton of mass Q_K into two partons close in mass instead of decay by on-shell bremsstrahlung, is a specific feature of the parton formfactor in QCD.

One gets

$$\frac{Q_{K+1}^2}{Q_K^2} \simeq \exp\left[-2\left[\ln(Q_K^2/Q_0^2)\right]^{1/2}\right] \tag{5.36}$$

One of the most important advantages of this scheme is that it allows to calculate the distribution of color sources in the final state. AMATI and VENEZIANO /115/ calculated the distribution of color among the n final partons of mass Q_0, with the astonishing result that already perturbative QCD provides a "preconfinement" of color. They discovered that these "final state" partons are grouped into colorless clusters in a number sufficient to "exhaust" the final state, but still possessing a finite average mass of the order Q_0^2. The result is peculiar of QCD, in particular of its nonabelian nature. Let me briefly sketch their derivation. In the axial gauge and at the leading log level, the flow of color during the degradation of a high Q^2 quark into partons at Q_0^2 is determined by planar diagrams, which means that color flow lines must not cross. Nonplanar diagrams are suppressed by $1/N_{color}$ factors.

This is visualized by Fig.5.4. Note that a quark corresponds to a single color line, whereas a gluon can be displayed as a color triplet-antitriplet state corresponding to two color lines. Obviously, it is always possible to group the "final state" partons into nonoverlapping, colorless clusters. Partons inside each cluster are nearby in the branching-tree structure of the jet. Consequently, the clusters have finite masses, as demonstrated in Fig.5.5.

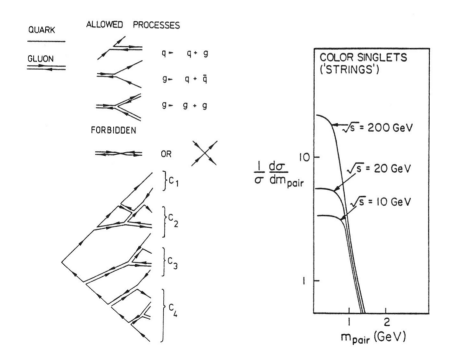

Fig.5.4. (a) Drawing rules for planar QCD graphs. (b) Planar graph of the process $e^+e^- \to$ hadronic clusters; the "final state" color lines can be grouped into color-singlet clusters $C_1 \dots C_N$

Fig.5.5. Mass distributions for color-singlet objects in $q\bar{q}$ jets at different energies. The distributions are calculated by a Monte Carlo simulation of the branching process, down to $Q_0^2 = 1$ GeV^2 /116/

The multiplicity of clusters is given by

$$\langle n_c \rangle \cong \exp\left[2\sqrt{\frac{N_c}{\pi b}} \ln(Q^2/\Lambda^2)\right] / \exp\left[2\sqrt{\frac{N_c}{\pi b}} \ln(Q_0^2/\Lambda^2)\right] \tag{5.37}$$

Notice that $\langle n_c \rangle$ grows faster with Q than any power of logarithms.

The importance of these results is obvious. Under the reasonable assumption that confinement at Q_0^2 converts these clusters into hadrons without a large reshuffling of color lines, the final state of the parton evolution is no longer a system of virtual colored objects which are hard to deal with. Instead, one has a number of physical objects, which can be treated as on-shell massive states and whose properties are independent of the initial Q.

There are now two possible ways of proceeding further. Either one chooses Q_0^2 as being relatively large, or $Q_0^2 \cong$ 10-100 GeV^2. The final-state, color-singlet clusters are then color flux strings connecting a quark and the corresponding antiquark. Their decay properties are known from e^+e^- annihilations at lower energies and are described by the phenomenological parton models described in Chap.4. The other way is to use a Q_0^2 as low as possible, O(1 GeV^2) /117/. In this case the final state clusters can be identified as the usual meson resonances and simple isotropic phase space volumes describes their decays with reasonable accuracy. This fragmentation model is once more illustrated in Fig.5.6. At the moment, however, it is hard to see how reliable quantitative calculations based on this last concept are, since the QCD expansion is used in a region where its convergence is questionable and where bound state corrections will be large.

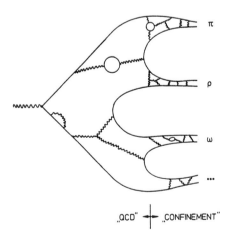

Fig.5.6. Perturbative and nonperturbative stages of parton confinement

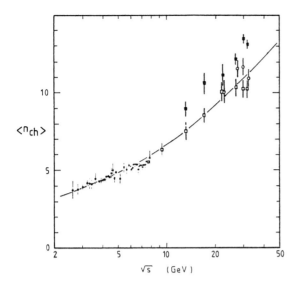

Fig.5.7. Comparison of measured and predicted hadron multiplicities. The theoretical curve shows the mean number of quanta at $Q_0^2 = 1$ GeV^2 in a $q\bar{q}$ jet and is normalized to data at $\sqrt{s} = 10$ GeV /118/

Nevertheless, the results turn out to be in fair agreement with data. Figure 5.7 shows the mean multiplicity of partons at $Q_0^2 \simeq 1$ GeV^2 in quark-antiquark jets as a function of the cms energy together with measurements. The model values are taken from Monte Carlo calculations /118/ where the parton shower is simulated in analogy to simulations of electromagnetic showers /119/, thereby keeping trace of all finite mass and finite transverse momentum effects. This last point has proven to be essential. The results differ appreciably from analytical calculations (5.37) which neglect such effects. In fact, asymptotic behavior is not yet reached at Q^2 as large as 10^4-10^5 GeV^2 /118/.

5.3 Production of Heavy Hadrons

Another essential aspect of scale breaking in quark jets is the production of heavy quark flavors. At sufficiently high Q^2, it is possible that a gluon radiated from the primary quark converts into a pair of heavy quark flavors, like charm. The charmed quarks fragment and produce charmed mesons.

In principle, such processes are contained in (5.27) and (5.30) since the p-functions (5.20) explicitly depend on the number of active flavors f. Practically these effects are neglected in most calculations because otherwise the p-functions get an additional Q^2 dependence

$$f = f(Q^2)$$

Neglecting threshold effects, the Q^2 dependence of the charmed meson production can be calculated from (5.30)

$$\frac{\partial}{\partial t} D(\alpha,t) = \frac{\alpha_S}{2\pi} D(\alpha,t) A\left[\alpha, f(t)\right] \quad \text{choosing } \ln\frac{m_c^2}{Q_0^2} \leq t \left(\leq \ln\frac{m_b^2}{Q_0^2}\right) \quad \text{and } f = 4$$

As a boundary condition

$$D\left(x, \ln\frac{m_c^2}{Q_0^2}\right) = \delta(1-x) \tag{5.38}$$

seems natural.

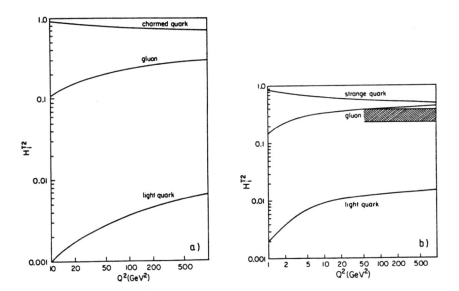

Fig.5.8. (a) The n = 2 moments of the total decay function into charmed hadrons, for charmed quarks, gluons and light quarks versus Q^2. (b) The n = 2 moments of the total decay function into strange hadrons, for strange quarks, gluons, and light quarks versus Q^2 /120/. Adding nonperturbative contributions all curves will be within the shadded area (for u-, d-, and s-quarks, the values given in Table 4.1 are shown; gluon will be between light and strange quarks)

Figure 5.8a shows the results for perturbative charm production by charmed quarks, gluons and light quarks as given by GEORGI and POLITZER /120/. The quantity plotted is the second moment of the charmed meson distribution, equivalent to the mean fraction of momentum carried by all charmed fragments. Qualitatively, the curves behave as expected, charmed initial quarks are most effective in producing charmed mesons. Only at high Q^2 additional quarks and gluons are produced perturbatively, thus diminishing the mean "charmed momentum". At first sight, however, the enormous difference between the light quark and gluon-fragmentation function seems surprising. The obvious reason is that a heavy quark can be produced by a primary gluon via pair production. In the light-quark case the gluon has to be taken from the steeply-falling bremsstrahlung spectrum.

Quantitatively, the model does not reproduce the presumably falling fragmentation functions of charmed quarks into a charmed meson /121/. This default is due to the boundary condition of (5.38) and it demonstrates that one has to be very careful in choosing these limits. The boundary conditions imposed on (5.30) have to be given in a region where the quark fragmentation is dominated by perturbative QCD effects and where the influence of phase space is negligible. Equation (5.38), however, is merely a consequence of the limited phase space available.

George and Glashow have as well tried to describe the perturbative production of strange mesons (Fig.5.8b) using $m_s \cong 500$ MeV. In this regime first-order perturbative QCD is not expected to hold, so the strong discrepancy to the measured values is not astonishing [using the FF jet as a parametrization of nonperturbative jets at moderate Q^2, one gets $D_s^S(2,Q^2 \cong 100) \cong 0.38$ and $D_{u,d}^S(2,Q^2 \cong 100) \cong 0.22$; gluons should be in between].

5.4 Transverse Momentum Structure of Parton Jets

In the naive quark-jet models, the lateral width of the jet is essentially determined by the diameter of the color string connecting the primary quarks. The mean transverse momentum of the produced particles is independent of the jet energy, Q^2, and of the x of the particles, apart from trivial kinematical effects at very low and at very high x. Such a naive jet represents a cylinder in momentum space.

All these statements become invalid as soon as one includes QCD scale breaking. This is easily visualized in the scale-invariant parton model of KOGUT and SUSSKIND (see Sect.5.1) /102/. Let us calculate the transverse momentum of a parton of size k-1 descending from a parton of size k. Since no other scale is present, the result must be proportional to the parton size

$$\langle p_\perp^2 \rangle_{k-1} \cong \left(\frac{1}{R_k}\right)^2 f(\alpha_k) \qquad (5.39)$$

where f is a function of the strong coupling constant α_k.

In asymptotic free theories one can calculate $f(\alpha_k)$ perturbatively. Only those "decays" where at least two partons of type k-1 are generated contribute to $\langle p_\perp^2 \rangle$. Therefore, one has $f(\alpha_k) \sim \alpha_k^2$. Summing up the transverse momenta squared of all steps of decay chain, one gets approximately for the $\langle p_\perp^2 \rangle$ of the "final state" partons

$$\langle p_{\perp p}^2 \rangle \sim Q^2 \alpha^2(Q^2) \sim \frac{Q^2}{\ln^2(Q^2/\Lambda^2)}$$

Furthermore, this smearing depends on the fraction of momentum carried by the hadron. A hadron at high x must contain partons from "early" stages of the decay and thus carry their large transverse momentum. A hadron at low x, on the other hand, will only contain one of the many fragments in the late stages of the decay and will receive only a small fraction of the early partons transverse momenta.

$$\langle p_\perp^2 \rangle^{1/2} \sim x \frac{Q}{\ln(Q/\Lambda)} \qquad (5.40)$$

A QCD jet at very high Q^2 is, therefore, represented by a cone in momentum space.

There are two ways to calculate the transverse structure of jets explicitly from QCD. First, it is easy to generalize (5.30) and include transverse momenta in the p functions /122/. Since in this case one has to know jet shapes at a reference Q^2, this method is not able to predict the width of QCD jets a priori. More precisely, as the estimates given above (5.39) and (5.40) in fact represent only lower limits, the method is not even able to prove the existence of finite-width QCD jets at all.

A second way, which avoids these problems, was proposed by STERMAN and WEINBERG /123/. They calculated the full QCD cross section in first order of α_s for quark-antiquark production by virtual photons. The diagrams contributing to σ are shown in Fig.5.9. As predicted by the KINOSHITA-LEE-NAUENBURG theorem /152/ all divergencies occuring in the single diagrams of Fig.5.9 cancel in the calculations of observable quantities.

As a measure of the transverse width of a jet, Sterman and Weinberg calculated the fraction f of events having a fraction ε of their energy outside a pair of cones of half angle $\delta \ll 1$. One obtains

$$f = 1 - \frac{4}{3\pi} \alpha_S(Q^2) \left[3\ln(\delta) + 4\ln(\delta)\ln(2\varepsilon) - \frac{\pi^2}{3} - \frac{7}{4} \right] \quad (5.41)$$

The values of ε and δ must be chosen such that $1 \gg (\alpha_S/\pi)\ln(\delta)\ln(2\varepsilon)$.

Fig.5.9. Diagrams contributing to the transverse width of QCD quark jets in first order of α_S /123/

Extensions of this formula to a wider range of parameters and calculations in higher orders have been published by several authors /125-127/. The main results of (5.41), however, remain unchanged: neglecting the effects due to confinement, QCD jets are well collimated in angle. At $\sqrt{s} = 7$ GeV one obtains 70% of all events having more than 80% of their energy within a double cone of half angle $13°$. At increasing energy and constant fractions, this angle decreases as $E^{-1/4}$. Thus, the conjecture that jets look narrower at higher energies is supported by QCD. The rate of shrinking of the opening angle is, however, much less than expected for fixed $p_\perp$ jets. At finite energies, (5.41) does not describe the final state hadrons. There, the nonperturbative smearing of transverse momenta has to be included. The interplay of the two components is demonstrated in Fig.5.10, which shows results of Monte Carlo calculations taken from /124/. In Fig.5.10 the half angle of a double cone containing 90% of the energy in a fraction f of all events is plotted as a function of Q^2. Dotted and dashed lines denote the contributions from the nonperturvative and perturbative components alone, respectively. One recognizes that below $Q^2 \simeq 1000$ GeV^2, QCD effects are small; above $Q^2 \simeq 10^4$-10^5 GeV^2, one has nearly pure QCD jets.

5.5 Gluon Jets

Up to now, we have dealt with the simplest type of jets, e.g., quark-antiquark jets resulting from the confinement of a color triplet-antitriplet system. However, color confinement is, of course, not restricted to color triplets; any attempt to separate nonsinglet color systems will produce jets. In QCD the second basic colored state besides the triplet is an octet, the gluon.

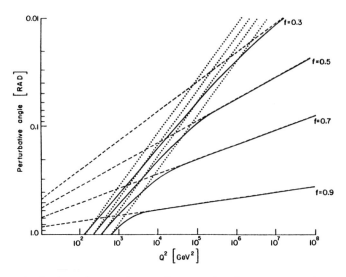

Fig.5.10. Half-opening angle of a double cone containing at least 90% of the total energy in a fraction f of all events of the reaction $e^+e^- \to$ hadrons. The dotted and dashed lines show nonperturbative and perturbative contributions, respectively /124/

Assume we generate two color octets of opposite color charge with a high invariant mass Q^2. This could happen in the decay of a heavy quark-antiquark bound state like the T into a photon and two gluons /128/. What will the jet look like?

The two most important predictions for hadron production by color octet sources are (referred to the two gluon cms):

I) the multiplicity in the rapidity plateau is increased by a factor 9/4 compared to quark jets /15,111/,

II) the color octet jet is softer, it contains less fast hadrons than a quark jet. Asymptotically one expects /15,110/

$$D_g^h(x,t) = (1-x)\, D_q^h(x,t) \qquad (5.24)$$
$$x \to 1$$
$$x \to \infty$$

Both features are already evident from the Schwinger model presented in Sect.4.2 Assume the primary partons have a color charge g' different from the coupling g which ties two fermions together to form a boson of the mass

$$m = \frac{g}{\sqrt{\pi}}$$

From (4.10) we now obtain

$$\frac{1}{\sigma}\frac{d\sigma}{dy} = \frac{1}{(2\pi)^2} \left| \int d\hat{x}\, e^{ipx} g\phi'_{ext} \right|^2 = \frac{g'^2}{\pi m} = \frac{g'^2}{g^2} \quad (5.43)$$

Now, remembering that in SU(N) the squares of charges of the N-plet and $N \times \bar{N}$-plet are given by the structure constants of the group, we obtain

$$g_N^2 = \frac{1}{N} \sum_a t^a t^a = \frac{N^2-1}{2N} = \frac{4}{3} \text{ for SU(3)}c$$

$$g_{N\times\bar{N}}^2 = \frac{1}{N^2-1} \sum_{abc} C_{abc} C_{abc} = N = 3 \text{ for SU(3)}c \quad (5.44)$$

So we expect

$$\left[\frac{1}{\sigma}\frac{d\sigma}{dy}\right]_{8\times\bar{8}} / \left[\frac{1}{\sigma}\frac{d\sigma}{dy}\right]_{3\times\bar{3}} = g'^2/g^2 = 9/4 \quad (5.45)$$

The relative softness of the gluon jet is now required by simple energy-momentum conservation.

In the QCD language these results are obtained by integrating (5.27)

$$-\frac{\partial}{\partial t} G(\alpha,t) = \frac{\alpha_S}{2\pi} A(\alpha) G(\alpha,t) \quad (5.46)$$

with boundary conditions corresponding to one gluon at $t = \ln(Q^2/Q_0^2)$. Asymptotically, the density of partons in the plateau of a gluon jet turns out to be 9/4 of the density in the quark plateau /110/. Correspondingly, the number of final state hadrons should be larger by roughly the same factor. It has been pointed out by BRODSKY that the last assumption may be only partially valid /129/. Due to the higher density of partons in rapidity, the average mass of the hadronic clusters formed in the last stage of confinement, as well as their decay multiplicity, should be smaller. Nevertheless, the height of the plateau of gluon jets should be roughly twice as large as in quark jets.

However, all these statements hold only at very large Q^2, where mass effects are negligible. Figure 5.11 demonstrates that with a finite Q^2, the difference between quark and gluon jets is smaller than predicted by (5.45).

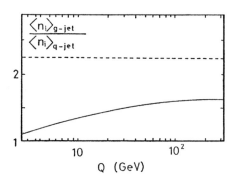

Fig.5.11. Ratio of parton multiplicities in gluon and quark jets for $Q_0^2 = 1$ GeV^2 as obtained from a Monte Carlo simulation of parton showers /118/. The dotted line shows the asymptotic limit

The hadron spectrum at high x from gluon jets can be estimated from the parton spectrum at $x \to 1$. In the limit of high Q^2, we get /110/ in analogy to (5.31):

$$q_g(x,0) \sim (1-x)^{16\xi/3}$$
$$x \to 1, \xi > 0.05$$
$$g_g(x,0) \sim (1-x)^{12\xi-1}$$
(5.47)

Equation (5.47) demonstrates the striking fact that the quark spectrum within a gluon is much flatter than the spectrum of gluons within a gluon jet. Thus the leading particle within a gluon jet at high Q^2 is a quark!

Since fast hadrons from gluon jets stem mostly from those quarks, we have

$$\frac{D_g^h(x,t)}{D_q^h(x,t)} \sim \frac{q_g(x,0)}{q_q^{ns}(x,0)} \underset{x \to 1}{\sim} (1-x)$$
(5.48)

The stronger coupling of gluons to gluons also leads to a different pattern of scaling violations. First, the mean perturbative opening angle of gluon jets is 9/4 of the opening angle of a quark jet at the same Q^2 /130,131/. Second, the effects of longitudinal scale breaking are increased by roughly the same factor. This is evident from Fig.5.12, where the Q^2 dependence of the nth moment of the final state hadron density (n = 2,4,6,8,10) is shown for gluon jets (a) and for quark jets (b) (taken from /132/).

In conclusion, using QCD we expect that gluon jets at high Q^2 have an increased decay multiplicity. The leading hadron spectrum is steeper by factor (1-x) compared to quark jets. Since gluons are flavor neutral, there is no "favored" leading flavor in a gluon jet, except perhaps mesons containing a strong glueball admixture.

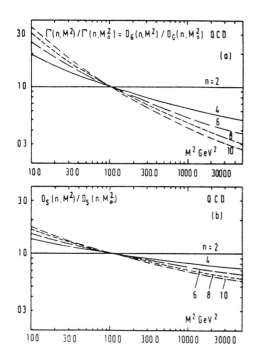

Fig.5.12a,b. Q^2 dependence of the nth moment of the fragmentation function of gluons (a) and quarks (b), normalized at $Q^2 = M^2 = 100$ GeV^2 /132/

5.6 Quantitative Test of QCD Predictions for Jets in e^+e^- Reactions

The main effect of QCD corrections in parton jets is a redistribution of the energy flux; longitudinal spectra are steepened and transverse spectra are widened.

In Chapt. 3 is shown that a scale breaking of longitudinal distributions could also be induced by nonasymptotic, nonperturbative effects. Furthermore, the expected size of QCD corrections to $d\sigma/dx$ is small (5.33) and beyond the accuracy of present measurements.

Therefore, quantitative tests of QCD corrections mainly deal with the transverse broadening of jets. Among the main questions to be answered are:

I) Is the effective coupling constant α_S, which governs scale breaking in quark jets, compatible with the values obtained in lN scattering at a similar Q^2?

II) Is gluon bremsstrahlung described by the QCD matrix element for spin - 1 boson emission?

III) How can the gluon fragmentation function be determined?

Quantitative results available up to now concern point I) and II).

As discussed in Chap.2, the increase in the mean $p_\perp$ of secondaries with respect to a fitted jet axis can be tracted to the appearance of events containing three planar jets. The number of such events with noncollinear jets, or, equivalently, the number of tracks in the tail of the $p_\perp$ distribution, is proportional to α_S.

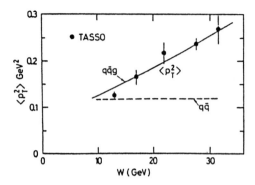

Fig.5.13. Mean transverse momentum squared with respect to the jet axis, as a function of $\sqrt{s}$ /24/ compared to QCD predictions /133/

Figure 5.13 proves that the increase of $\langle p_\perp^2 \rangle$ is well accounted for by QCD predictions.

Perturbative and nonperturbative components of $\langle p_\perp^2 \rangle$ can be separated when transverse momenta in and out of the "event plane" are considered. The momenta out of the plane are determined by nonperturbative processes, whereas momenta in the event plane also receive contributions from a third jet emitted at large angles. Figure 5.14a shows the distribution of the mean $\langle p_\perp^2 \rangle_{IN}$ and $\langle p_\perp^2 \rangle_{OUT}$ per event /24/. Also indicated are predictions for different values of the scale parameter Λ in QCD /134/. The range of parameters compatible with the data is given in Fig.5.14b. Data are fully consistent with $\Lambda \simeq 0.5$ GeV, as determined from lN scattering /108/, and with a constant nonperturbative transverse momentum of the order 300 MeV, as obtained for jets at lower energies.

Similar conclusions are obtained by studying the "oblateness" of the distribution of transverse momentum vectors in jets /42/, or simply by counting the number of noncylindrical planar events (Fig.5.14c).

First evidence for a spin 1 nature of the gluons observed in e^+e^- scaling violations comes from a recent work of the TASSO group where the distribution of the emission angle of the third jet is investigated (Fig.5.15) /136/.

5.7 Summary

At large q^2, where α_S is small, the evolution of jets, as discussed in Chap.4 in a very phenomenological way, can be calculated perturbatively in QCD. The characteristic feature of these "QCD jets" is a broadening of the transverse size of a jet proportional to $Q^2/\ln(Q^2)$ and a steepening of longitudinal distributions. The QCD contributions to the transverse jet size should be relevant at $Q^2 \simeq 1000$ GeV^2. The effects seen at highest PETRA energies are compatible with QCD predictions, both qualitatively and quantitatively.

The QCD model can be applied also to gluon fragmentation. The main predictions concerning gluon jets are:

- the longitudinal distribution of fragments is softer than in quark jets, the opening angle of a gluon jet is larger;
- the production of heavy flavors is enhanced.

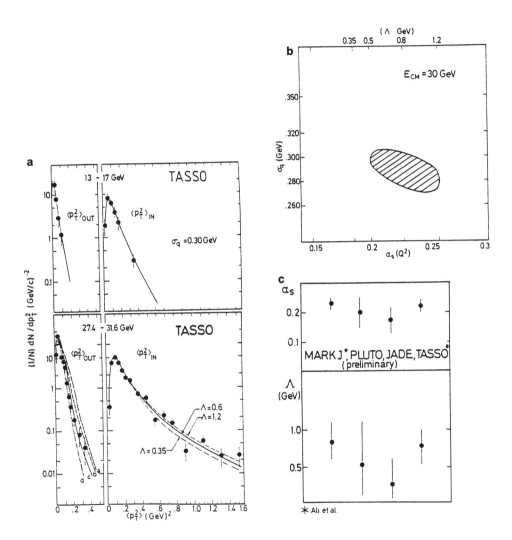

Fig.5.14a-c. (a) Distribution of the mean momentum squared in and out of the event plane at $\sqrt{s} \simeq$ 13-17 GeV and at $\sqrt{s} \simeq$ 30 GeV /24/ as compared to QCD predictions /134/, assuming an average nonperturbative transverse momentum of fragments σ_q = 300 MeV. (b) Allowed domain for the parameter σ_q and the QCD scale parameter Λ, for the data shown in (a). (c) Summary of values of α_s as determined from experiments at PETRA at $\sqrt{s} \simeq$ 30 GeV /24,43,134,135/

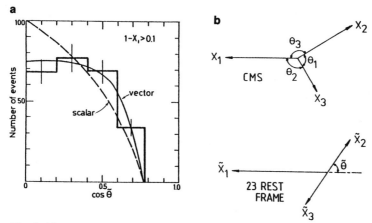

Fig.5.15a,b. Distribution of the emission angle $\tilde{\theta}$ for a third jet in planar events in e^+e^- annihilations at $\sqrt{s} \simeq 30$ GeV compared to predictions for spin 1 and spin 0 gluons /136/. The angle $\tilde{\theta}$ is measured in the rest system of the two slower partons 2 and 3, as evident from (b)

6. The Fragmentation of Parton Systems

In Chap.2 and 4, we considered interactions where a four momentum q is transferred to a single parton giving rise to the production of jets. We restricted our attention to very simple initial conditions. The systems studied consisted of a parton and an antiparton of opposite color. Furthermore, only elementary, irreducible partons were considered, i.e., quarks and gluons. These limitations reduce the number of variables. The resolution power $1/\lambda \cong \sqrt{|q^2|}$ of the probe is identical to the invariant mass $\sqrt{s}$ of the hadronic final state. In the picture of color flux strings, the geometry of the color field connecting the two partons is well defined. Experimentally, these conditions are fulfilled only in one type of reaction; the decay of a heavy (timelike) state into quark or gluon pairs. Other deep inelastic reactions, like the scattering of quarks out of a nucleon by probes of large spacelike momenta, create more complex final states which cannot be fully described by the tools discussed above. In this chapter we shall try to give a phenomenological description of these reactions.

6.1. Deep Inelastic Lepton-Nucleon Scattering

As a first example let us study processes where the nucleon structure is probed by a photon or a weak vector boson of high spacelike momentum. Various interactions are possible. In the simplest case, the probe is absorbed by a valence quark. Figure 6.1 compares the corresponding diagrams of parton and color flux lines with those obtained for quark production in e^+e^- annihilations. The main difference of the parton final state is that in Fig.6.1b the antiquark is replaced by the diquark system ("spectator") remaining after the "active" quark is scattered. The diquark forms a color antitriplet

$$3 \times 3 = \bar{3} + 6 \tag{6.1}$$

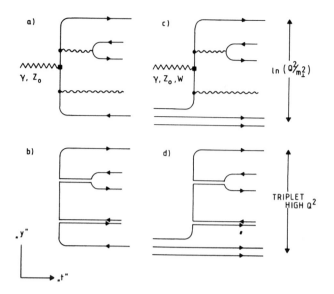

Fig.6.1a-d. Schematic representation of hadron production in e^+e^- reactions (a,b) and in lepton-valence quark scattering. Parts (b) and (d) show the planar diagrams of color flux corresponding to (a) and (c)

The sextet state is excluded because the whole system has to be a color singlet. The (transverse) mass of the spectator diquark is determined by the primordial (transverse) mass of the active quark, $m_{\perp q} = (m_q^2 + p_\perp^2)^{1/2}$, and by its momentum fraction x

$$m_{\perp qq}^2 \simeq (1-x)(m_n^2 - m_{\perp q}^2/x) \qquad (6.2)$$

For x not too small, the diquark invariant mass is of the order of the nucleon mass m_n. Thus it seems natural to assume that they still form a coherent state /137/.

Also indicated in Fig.6.1 are radiative corrections leading to preconfinement. The diagrams are identical for e^+e^- annihilation and lepton-nucleon scattering. The physical difference is that in the e^+e^- case, quark and antiquark radiate after the hard process, whereas in the latter case, the quark radiates immediately before and after the hard scattering.

Due to the analogy of diagrams, the hadron distribution in the final state (Fig. 6.2b) should be identical to that observed in e^+e^- annihilations, except that the antiquark fragmentation region is replaced by a diquark fragmentation region.

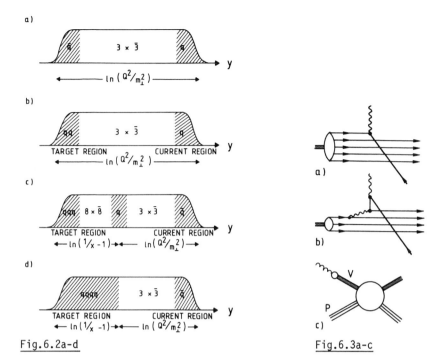

Fig.6.2a-d

Fig.6.3a-c

Fig.6.2a-d. Structure of the hadronic final state (a) in e^+e^- annihilations; (b) in lepton-valence quark scattering; (c) in lepton-sea quark interactions resulting in an incoherent spectator; (d) in lepton-sea quark interactions resulting in a coherent spectator. In lepton-nucleon interactions the rapidity axis is defined by the three momentum of the current

Fig.6.3a-c. Possible interactions between a spacelike current and nonvalence partons. (a) Scattering off a sea quark belonging to the primordial hadron wave function; (b) scattering off a sea quark created dynamically during the interaction; (c) interaction with the vector meson component of the current by soft exchange of wee partons

Another possible interaction is the interaction of the "pointlike" current with a sea parton belonging to the primordial Fock state of the nucleon (Fig.6.3a). Then the final state hadrons populate a rapidity region of the size

$$Y = \ln(W^2/m_\perp^2) \qquad (6.3)$$

The invariant mass W of the hadronic final state is given by

$$W^2 = (p_n+q)^2 = Q^2(1/x-1)+m_n^2 \qquad (6.4)$$

$m_\perp$ is a typical transverse mass of hadrons in jets, $m_\perp \cong 0$ (0.5 GeV). The range Y is exhausted by two types of fragments: those arising from the confinement process of the active quark are spread over a region $Y_q \cong \ln(Q^2/m_\perp^2)$, like in e^+e^- annihilations. The residual region $Y_T \cong \ln(1/x-1)$ is populated by spectator fragments (Fig. 6.2c) /137-143/.

Details of the structure of the final state can be derived by considering the configuration of color sources. For definiteness let us use rapidities in the target rest frame and assume that an antiquark has been struck. The scattered quark represents an antitriplet at $y_{LAB} \cong Y$, the opposite color pole being the spectator at $y_{LAB} \lesssim \ln(1/x-1)$. In analogy to e^+e^- reactions, we thus expect to see a plateau region of the length $\cong Y-\ln(1/x-1)-2...3$ followed by a shoulder at $y_{LAB} \cong Y$ containing the direct fragments of the active quark.

Predictions concerning the region of the target or spectator fragments are less evident. The standard assumption /138,139,143/ is the following: quark-antiquark pairs from the sea have low invariant masses and the rapidities of quark and antiquark are similar. So the sea quark left over is localized at the primordial rapidity of the active quark, $y_{LAB} \cong \ln(1/x-1)$. In analogy to the Dirac picture of antiparticles as hole states in a neutral continuum, this quark is often referred to as the quark hole. At $y_{LAB} \cong 0$, the three valence quarks form a color octet since much earlier one of them emitted the gluon which created the sea quark pair. Consequently, one may expect to find the valence and hole fragments at $y_{LAB} \cong 0$ and at $y_{LAB} \cong \ln(1/x-1)$, respectively, connected by a hadronic plateau created by the octet color field.

It was pointed out by BRODSKY /137/ that this model may not be appropriate. The initial five quark Fock state is a soft, coherent, and long-lived fluctuation of the nucleon core. In a coherent state, however, it is impossible to further localize color charges.

This is easily demonstrated in the parton picture. The relatively small mass of a nucleon implies that the mean transverse mass of sea quarks goes to zero as x goes to zero /144/. Then, however, the rapidity spread of a low mass $q\bar{q}$ pair may be arbitrarily large.

Under these conditions, any further subdivision of the target region seems unmotivated (Fig.6.2d). We shall return to this point later when discussing specific models of the fragmentation of multiquark states.

Besides the scattering off of a primordial sea parton, a quark pair may be created dynamically as an integral part of the scattering process due to the pointlike coupling of the current to a quark line (Fig.6.3b). In this case, the residual quark is off shell of the order $0(Q^2)$ and is localizable within the target nucleon. Because of its large distance of the mass shell and its short lifetime, this quark hole is an incoherent part of the spectator wave function and fragments independently. Through the exchange of hard quanta, even the coherence of the nucleon's valence core may be partly destroyed. This process should dominate lepton-nucleon interactions at high Q^2; the resulting final state will resemble (Fig.6.2c).

Of course, the strict division between coherent and incoherent sea partons, between primordial quarks and those created dynamically, and between nonperturbative and perturbative effects are somewhat arbitrary. In reality one will have a smooth transition between the two regimes.

At low Q^2, $Q^2 \lesssim 0$ (1 GeV^2), coherent soft interactions between current and target compete with the above hard processes. At low Q^2 and W^2 quasi-elastic scattering of the lepton may occur, resulting in the production of excited nucleon states. At higher W, the vector meson component of weak or electromagnetic currents may interact with the nucleon by exchanging wee partons (Fig.6.3c). Both target and current region form coherent states.

Obviously such interactions are not suited to study parton jets. In the experimental data quoted in the following discussion, quasi-elastic events are removed by cuts in Q^2 and in W (typically $Q^2 > 2$ GeV^2, W > 4 GeV). The cut in Q^2 as well suppresses coherent interactions. For almost all applications, the requirement $Q^2 > 2$ GeV^2 is selective enough.

Before going into a more detailed discussion, a few words concerning the choice of reference frames and variables.

The three-momentum component of the current defines a preferred axis along which the color sources involved are more or less aligned. This fixes the reference frame except for longitudinal boosts. Four conventions are usual:

- the laboratory frame in which the target was at rest before the interaction. The usual scale variable for jet studies is $z = E_{hadron}/$total hadronic energy E_H.
- The center of mass frame of the hadronic final state. Scaling variable is $x_F = 2p_{\shortparallel}^*/W$.
- The Breit frame, in which the probe carries only momentum, but no energy. Scaling variable is $z_B = 2p_{\shortparallel}^B/|q|$.
- The rest frame of the active quark after scattering.

Neglecting nucleon and quark masses, the frames are connected by the boosts (Fig.6.4) (with Q measured in ~GeV)

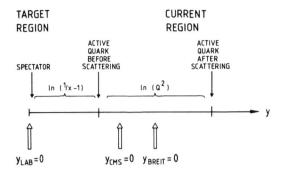

Fig.6.4. Interrelation of the various reference frames used to describe the hadronic final state in lepton-nucleon interactions

$$y_{CMS} = y_{LAB} + \ln[Q^2(1/x-1)]^{1/2}$$
$$y_{BREIT} = y_{LAB} + \ln[Q(1/x-1)] \qquad (6.5)$$
$$y_{QUARK} = y_{LAB} + \ln[Q^2(1/x-1)]$$

Obviously, the laboratory and the quark rest frame are suited to study the fragmentation of the valence core of the target, and of the scattered quark, respectively. The center of the color triplet plateau is characterized by $y_{BREIT} = 0$.

The scaling variables quoted above are used to describe the fragmentation of the scattered quark. For fast fragments, $p_\parallel \gg m_\perp$, the variables agree asymptotically.

The study of interactions of pointlike probes with hadrons provides the opportunity to check our ideas on quark fragmentation in a more complex environment compared to e^+e^- annihilations, and it further provides information on the fragmentation of compound parton systems, such as diquarks.

To arrive at a unified description of the various hard processes and to enable quantitative predictions, we shall discuss the principles of scaling, environmental independence, factorization, and jet universality.

6.2. Environmental Independence and Factorization

The hypothesis of environmental independence is implicitly contained in the genuine quark-parton model /1-4,80/ and was first stated explicitly by SEGHAL /90/. Environmental independence means that the distributions of hadrons in the parton fragmentation region of a specific process is expressible as

$$\frac{1}{\sigma}\frac{d\sigma^h}{dz} = \sum_i \epsilon_i D_i^h(z) \qquad (6.6)$$

where ϵ_i is the probability that the fragmentation parton is of type i ($\sum_i \epsilon_i = 1$) and $D_i^h(z)$ is the fragmentation function depending on a scaling variable z. This means that for the purpose of calculation, we may assume that the primary interaction creates a "parton beam" with a composition given by the probabilities ϵ_i, and that each parton variety i decays into hadrons of type h with a distribution $D_i^h(z)$ which is independent of the origin of the parton.

The probabilities ϵ_i characterize the process under investigation. In e^+e^- annihilations the ϵ_i are proportional to the quark charges squared, whereas in lepton-nucleon interactions they further depend on the fractional momentum x of the interacting parton.

In (6.6) the fragmentation functions are assumed to scale. The scaling variable used by Seghal to describe quark fragmentation in lepton-nucleon reactions was z, defined as the ratio of the actual and the maximum hadron energy in the laboratory frame. In general, the choice of the "suited" reference frame is ambiguous. Figs. 6.2 and 6.3 suggest using z_B in the above case, whereas naively one may choose x_F.

This problem disappears at infinitely high energies. Here the relation

$$p_h = zp_q \qquad (6.7)$$

between the four momenta p_h and p_q of quark and hadron, respectively, holds in all frames with the same z. At finite energies (6.6) should hold for those hadrons which are fast (E >> m_h) in each of the various frames.

What is the experimental evidence for factorization? We have seen in Chap.4 that the fragmentation functions measured in e^+e^- annihilations, in electroproduction, and in neutrino-nucleon interactions can be described by one unique parametrization. This is once more demonstrated in Fig.6.5, where the z distribution of charged hadrons is summarized for the various reactions /93/. The measurements agree at not too high z within the error bars. At z > 0.8 a slight discrepancy seems to develop. One should bear in mind, however, that the distributions have been measured using drastically different detectors; that acceptance corrections and selection criteria (e.g., on the minimum number of reconstructed tracks per event) have been applied; and last, but not least, that the Q^2 range and the final state masses differ with typical Q^2 being as low as 2 GeV^2 in the neutrino reactions compared to $Q^2 \simeq 0$ (25 GeV^2) in the e^+e^- case. All these differences mainly influence the region close to z ≅ 1.

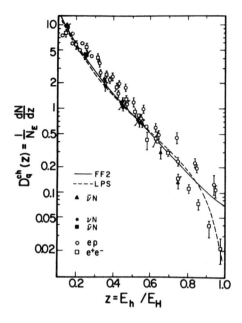

Fig.6.5. The fragmentation $D_q^{h^\pm}(z)$ for charged hadrons in various hard reactions /93/

In all types of quark jets, typical transverse momenta with respect to the axis turn out to be of the order 300-400 MeV/c, and all show the seagull effect /145/.

On the other hand, do we expect exact environmental independence and scaling for quark jets? It has been pointed out that phase space effects lead to nonscaling of fragmentation functions even if the basic matrix element is scale invariant. Thus we get

$$D_i^h(z) \xrightarrow[\text{phase space}]{} D_i^h(z,W) \qquad (6.8)$$

Following the argumentation of Kogut and Susskind (see Chap.5), a scattered parton is not fully described by the parton type; one has to know as well its "size" Q^2.

$$D_i^h(z) \xrightarrow[\text{field theory}]{} D_i^h(z,Q^2) \qquad (6.9)$$

The scale breaking leads to a shrinking of the z distribution and to a broadening of $p_\perp$ distribution within the jet.

Moreover, we have seen in Sect.6.1 and in Chap.5 that the quark composition of the target, as probed by a pointlike current, depends on q^2 as well, and that the clean separation of target structure and interaction is no longer possible. Does this imply the total breakdown of the idea of environmental independence?

Fortunately, QCD provides us with the equivalent principle of factorization, as far as the Q^2 dependence of the process is concerned. Fig.6.6a describes the interaction of a probe with a hadron as seen in the naive quark model. The incoming nucleon splits into a quark and a spectator. The quark scatters like a pointlike object, and both active quark and spectator decay independently. The QCD version of the process is shown in Fig.6.6b. The incoming hadron dissociates into a primordial quark and a spectator according to the structure function $G(x')$. Prior to the interaction with the probe, the quark radiates and becomes off shell of order Q^2. It absorbs the probe and again radiates quanta to approach again the mass shell. The practically on-shell quark and the spectator decay with fragmentation functions $D(z')$, where z' is the ratio of the hadron and the dressed quark energies. Despite these effects, environmental independence should still hold since it has been proven that the violations of scaling induced by radiation before and after the hard scattering processes are independent of the precise nature of the hard process /146-148/. The naive model of scale breaking by Kogut and Susskind also suggests such a behavior.

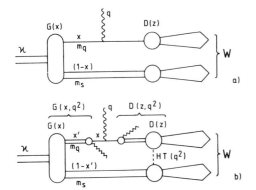

Fig.6.6a,b. Lepton-nucleon interactions seen at various levels of complexity. (a) Naive quark model. QCD corrections are indicated in (b). Double lines refer to the propagators of hadrons or dressed partons; "pointlike" partons are represented by single lines

The scaling violations can be absorbed in the definition of the now Q^2 dependent structure and fragmentation function. In lepton-nucleon scattering, the cross section factorizes as

$$\frac{d\sigma}{dxdz} = \frac{d\sigma}{dx}(x,Q^2)D(z,Q^2) \qquad (6.10)$$

The factorization was shown to hold for the leading QCD corrections $[\sim \alpha_s(Q^2)\ln(Q^2)]$ and in the leading log approximation /146/. So factorization is the QCD analogy of the naive parton model's environmental independence.

Recent investigations have shown that factorization, however, is only approximated in QCD and is violated by terms $O(\alpha_S)$ /149-151/. This is demonstrated by Fig.6.7 where the effective quark fragmentation functions as calculated by QCD up to $O(\alpha_S)$ for deep inelastic scattering processes are compared for different x /151/. The fragmentation functions depend on x as well as on z, the violations of scaling being strongest at large x and z. Although the effects of nonfactorization are small in those regions where the bulk of present data lies, one should be aware that this effect signals the point were the intuitive parton picture breaks down; it no longer makes sense to treat quark distributions and fragmentation functions as separate entities.

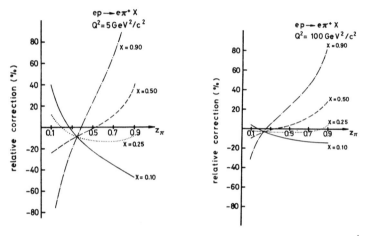

Fig.6.7. Relative correction to the fragmentation function D_q^h of order $\alpha_S(Q^2)$ for the reaction $ep \to e\pi^+ + x$ as a function of z for different values of the Bjorken scaling variable x at $Q^2 = 5$ GeV^2 and at $Q^2 = 100$ GeV^2 /151/

Besides these corrections related to the emission of quanta at large Q^2, it is still a controversy to what extent bound state effects modify the asymptotic predictions /153-156,158/. Apart from "trivial" target mass corrections /157/, the exchange of soft gluons between an active quark and a spectator may lead to additional corrections depending both on x and Q^2. Although the importance of these "higher twist" effects is proportional to powers of the interaction time $\tau \sim 1/Q$ /155-157/ (and is therefore negligible at $Q^2 \gg m_{nucleon}^2$), they may yield large corrections at moderate Q^2 /159/.

What is the experimental situation? From Fig.6.5 we learned that factorization at least approximates reality. However, experimental uncertainties make a definite conclusion difficult. A better test of factorization is possible in lepton-nucleon scattering where $d^2\sigma/dxdz$ can be measured for various x using the identical detector and acceptance corrections. Even in this case, however, the interpretation of data is not unambiguous. Because of

$$W^2 = Q^2(1/x-1) \tag{6.11}$$

any variation of x with Q^2 fixed (to test factorization) or of Q^2 with x fixed [to measure $D(z,Q^2)$ to compare with QCD predictions] changes W as well, giving rise to scale breaking due to phase effects.

Figure 6.8 shows the distribution of masses W of the hadronic final state as measured by a typical neutrino-nucleon scattering experiment performed at the Fermilab wide band beam using the 15 ft hydrogen bubble chamber /160/. The bulk of data lies below W < 5 GeV. In e^+e^- reactions, however, this is the point were a jet structure starts to be visible, averaged over many events! In that region, the structure of a single event is dominated by phase space effects. This is once more visualized in Fig.6.9 where the mean multiplicities in the quark fragmentation region z > 0.2 are plotted vs W /160/. Scaling implies that these multiplicities should be constant. The deviations are due to threshold effects and are well explained by a phase space model (solid lines). Using a cut W > 4 GeV to exclude the influence of phase space limitations, the experimentalists claim to see no significant violations of scaling or factorization.

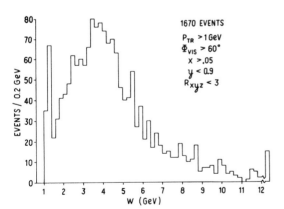
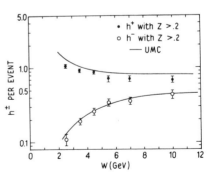
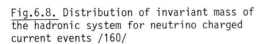

Fig.6.8. Distribution of invariant mass of the hadronic system for neutrino charged current events /160/

Fig.6.9. The rate of production of positive and negative hadrons with z > 0.2 as a function of W for neutrino-proton interactions. The curves are the result of a longitudinal phase space model /160/

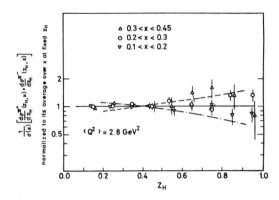

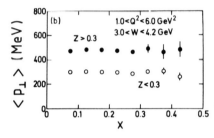

Fig.6.10. The cross section $d\sigma/dxdz$ for electroproduction normalized to its average over x at fixed z /161,162/. The curves illustrate the expected violation of factorization, calculated in $O(\alpha_S)$ /151/, for x = 0.15 (−·−·−), x = 0.25 (———), and x = 0.37 (-----)

Fig.6.11. Mean transverse momentum of hadrons in electroproduction as a function of x /161, 162/

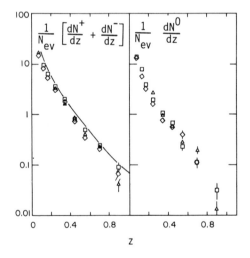

0.1 < x < 0.2

□ 10 GeV^2 < Q^2 < 20 GeV^2
△ 20 GeV^2 < Q^2 < 40 GeV^2
◇ 40 GeV^2 < Q^2

Fig.6.12. Quark fragmentation functions into charged and neutral hadrons for three bins of Q^2 as derived from muoproduction /163/. The full line shows the prediction of the FF jet model for $Q^2 \sim 10$ GeV^2

The same result was obtained with higher statistics by an electroproduction experiment at CORNELL /161,162/. Figure 6.10 shows $d^2\sigma/dxdz$ for various x, normalized to the average $d\sigma/dz$. Cuts on the four-momentum transfer $Q^2 > 2$ GeV^2 and on $W > 3$ GeV suppress kinematical effects at low masses and reject quasi-elastic events. Within the x range covered, no significant violations of factorization are seen, the upper limits being still compatible with QCD predictions (Fig.6.7). Figure 6.11 shows that at fixed W the mean transverse momentum of fragments is independent of x as well.

The same conclusions were obtained by an experiment studying $\bar{\nu}p$ interactions /167/ and by a recent muoproduction experiment investigating the region of large Q^2, 5 GeV$^2 \leq Q^2 \leq 100$ GeV^2. Within the experimental accuracy, no violations of scaling or factorization were obtained (Fig.6.12) /163/.

A controversial result was reached by a recent high-statistics bubble-chamber experiment on νp interactions /164/. Figure 6.13 shows the inclusive distribution $(1/\sigma)(d\sigma/dz)$ for two ranges of Q^2 averaged over all W. Significant differences are observed when going from $Q^2 \simeq 1-2$ GeV^2 to $Q^2 \simeq 5-40$ GeV^2. Because of the interrelationship of x, Q^2 and W this signals either a breakdown of factorization or of scaling, or the dominance of phase space effects. To investigate this in more detail, the moments

$$D(n,Q^2) = \int_0^1 z^{n-1} D(z,q^2)dz \qquad (6.12)$$

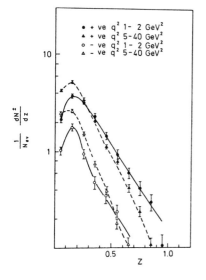

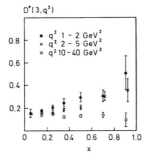

Fig. 6.13. z distributions ($z = E_h/E_{MAX}$) of positive and negative hadrons for two ranges of low and high Q^2 from BEBC νH_2 /164/ for hadrons going forward in the cms ($x_F > 0$)

Fig.6.14. Third fragmentation moment of positive hadrons with $x_F > 0$ vs Bjorken x for three intervals of Q^2. From BEBC νH_2 /164/

have been calculated and compared to QCD predictions /149/. Figure 6.14 shows the x dependence of the third moment of the density of positive particles in the quark fragmentation region $x_F > 0$ for three intervals of Q^2. For $Q^2 \simeq 1$ to 5 GeV^2, a clear violation of factorization is observed; $D^+(3,Q^2)$ depends strongly on x at fixed average Q^2. On the other hand, the moments of the fragmentation function depend strongly on Q^2 as well, when averaged over x (Fig.6.15). The observed nonscaling is in rough agreement with the QCD predictions based on (5.33), although the fit yields a rather large scale parameter $\Lambda \simeq 0.6$ GeV. Note however, that the final state mass W ranges between 1 and 3 GeV for $x > 0.3$ and $Q^2 = 2$ GeV^2! It would be rather astonishing if factorization or scaling, in the sense of QCD, holds at these masses.

In fact, if the moments are investigated at fixed W, i.e., if a cut $W > 4$ GeV is used like in most other experiments, data agree with scaling and factorization; the Q^2 dependence of the moments disappears completely (Fig.6.15b).

It is possible, that the good agreement with QCD, as obtained in Fig.6.15a by fitting only one free parameter Λ, is faked by phase space effects?

This question has been investigated by ENGELS et al. /165/ by using the example of jets in e^+e^- annihilations. They compared the shape of $D(n,Q^2)$ as predicted by QCD, with scaling violations given by a longitudinal phase space model as described in Chap.3 (Fig.6.16). The shapes of the two sets of curves are very similar; at low Q^2 or S, the phase space model predicts an even stronger variation of $D(n,Q^2)$ with Q^2. So this model would even explain why no violations of scaling are observed above $W = 4$, whereas in QCD one expects scaling violations to be independent of W, at least if $O(\alpha_s)$ correlations are neglected.

To summarize so far, the violations of scaling and factorization observed in longitudinal distributions in quark jets are compatible with kinematical threshold effects which vanish above $W \simeq 4$ GeV. Such a behavior is, within the limited statistics of data, still consistent with QCD, since QCD corrections to longitudinal fragmentation functions are relatively small [see (5.33)].

We have seen in Chap.2 that the transverse size of jets exhibits nonscaling at the highest PETRA energies. Nevertheless, in most cases only tails of distributions are concerned, and a constant mean $p_\perp$ for fragmentation is still a fairly reasonable assumption.

Thus, we conclude that at present energies, the principles of scaling and factorization provide a good description of the main features of quark jet systems.

A direct consequence of factorization is the retention of quark quantum numbers at high rapidities of the current fragmentation region, up to a constant term. This has been demonstrated qualitatively in Chap.4; here we will briefly summarize quantitative results.

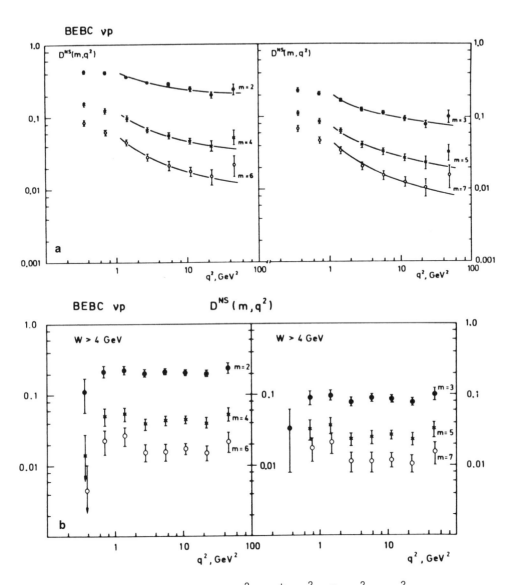

Fig.6.15a,b. Nonsinglet moments $D_{ns}(n,Q^2) = D^+(n,Q^2)-D^-(n,Q^2)$ vs Q^2 for m = 2-7 /164/. (a) Averaged over all W. The curves show fit of the QCD formula (5.33) to the data points above $Q^2 = 1$ GeV^2; (b) for W > 4 GeV

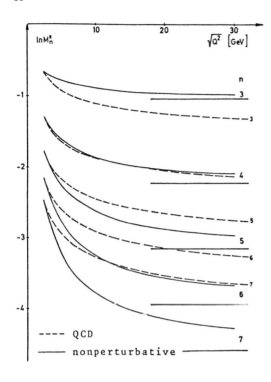

Fig.6.16. Q^2 dependence of the logarithms of moments from nonperturbative quark fragmentation with respect to the scaling variable x_F compared to perturbative QCD prediction calculated with $\Lambda^2 = 0.5$ GeV^2. Horizontal lines indicate the scaling limits of the nonperturbative moments /165/

Clean "beams" of one quark species can be produced in the following reactions:

a) *neutrinoproduction* $\nu N \to l^- + u + x$
$\bar{\nu} N \to l^+ + d + x$
for $x > 0.1$

b) *electroproduction* $lp \to l + u + x$
for $x \to 1$

where l refers to an electron or muon. The cuts in x are necessary to suppress interactions with sea quarks in a). The limit $x \to 1$ in b) makes use of the fact that for $x \to 1$ a proton consists of pure u-quarks, in addition to the fact that the u-quark cross section is enhanced by a factor 4 compared to d-quarks because of its charge.

The main experimental limitations of the study of quantum number retention comes again from the low masses of the hadronic final state which induce a considerable spillover between the target and the current fragmentation region. This is evident from Fig.6.17. The total negative charge in the d-quark fragmentation region in $\bar{\nu} N$ reactions increases with W, since then the contamination due to positive fragments of the target nucleon becomes less important /166/. The real mean charge of quark fragments was derived from Fig.6.17 by extrapolation linear in W^{-1} to $W = \infty$ (Fig. 6.18). The result, $\langle Q_d \rangle = -(0.46 \pm 0.8)$, is in good agreement with the expectation

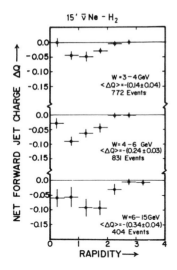

Fig.6.17. Distribution of charge per event in the cms rapidity for hadrons going forward in the cms in events with x > 0.1 and $Q^2 > 1$ GeV^2 for three intervals of W. From 15' $\bar{\nu}$Ne-H$_2$ /166/

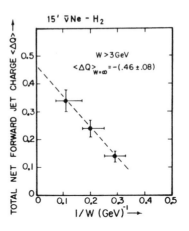

Fig.6.18. Total net charge <Q> per event of hadrons going forward in the cms vs 1/W. From 15' $\bar{\nu}$Ne-H$_2$ /166/

Table 6.1. Measured net charges in the u,d fragmentation region

Experiment	Ref.	Selection criteria and cuts (in units of GeV, GeV^2 etc.)	$<Q_u>$	$<Q_d>$
15' $\bar{\nu}$H$_2$Ne	/166/	$W \to \infty$, $x > 0.1$, $Q^2 > 1$, $x_F > 0$		-0.46 ± 0.08
BEBC νH$_2$	/164/	$W \to \infty$, $x > 0.1$, $Q^2 > 1$, $x_F > 0$	0.59 ± 0.10	
BEBC νH$_2$	/164/	$W \to \infty$, $x > 0.1$, $Q^2 > 1$, $Z_B > 0$	0.52 ± 0.08	
BEBC $\nu, \bar{\nu}$NeH$_2$	/168/	$W > 4$, $x > 0.1$, $Q^2 > 1$, $Z_B > 0$	0.55 ± 0.06	-0.12 ± 0.08
BEBC $\nu, \bar{\nu}$NeH$_2$	/168/	$W > \infty$, $x > 0.1$, $Q^2 > 1$, $Z_B > 0$		-0.3 ± 0.1
DECO ep	/161/	$W > 2.5$, $x \to 1$, $Q^2 > 1$, $x_F > 0$	0.48 ± 0.05	

based on the FF jet model. The experimental information on charge retention in quark jets is summarized in Table 6.1.

In the high energy limit, all data are in qualitative agreement with the expectation of approximate charge retention. Quantitatively, the data tend to confirm an SU(3) breaking for sea quarks as used in the FF model.

6.3 Jet Universality

At moderate energies the jet physics is dominated by the parton fragmentation regions investigated in the last section. At asymptotic energies, however, the direct quark or spectator fragments populate only a relatively small region of the whole rapidity range; the dominant contribution to particle production comes from the plateau region. The idea of jet universality tries to relate the structure of the rapidity plateaus observed in the various deep inelastic reactions.

The idea of an "universal" plateau, whose properties are asymptotically independent of target and current in current-induced reactions was first suggested by BJORKEN and KOGUT /169/ based on correspondence principles. In an universal plateau the density of each particle species $(1/\sigma)(d^2\sigma/dydp_\perp^2)$ is independent of the final state mass and the reaction considered. Consequently, one predicts an universal behavior of multiplicities.

$$<n> = n_0 \ln(W^2) \tag{6.13}$$

Taking into account our present ideas on parton fragmentation, this model has to be modified slightly. In two-dimensional quark-gluon bremsstrahlung models, as discussed in Chaps.4 and 5, the mean multiplicity per unit of rapidity grows with the square of the partons color charge. This observation led BRODSKY and GUNION /15/ to the following universality principle: the properties of the hadronic plateau are uniquely determined by the type of color charges separated, independent of the flavor content of the color sources.

In QCD, the two basic sources of color are triplets, such as quarks or diquarks, and octets, such as gluons or quark-antiquark pairs.

In current-nucleon interactions at high Q^2, one should observe a triplet plateau of the length $Y_{3\times\bar{3}} \simeq \ln(Q^2)$ and on octet plateau of the length $Y_{8\times\bar{8}} \simeq \ln(1/x-1)$ (Fig.6.2c), thus, yielding the asymptotic multiplicity

$$<n> = n_{3\times\bar{3}}\ln(Q^2) + n_{8\times\bar{8}}\ln(1/x-1) \atop {x \to 0 \atop Q^2 \to \infty} \tag{6.14}$$

More generally, and taking into account scaling violations, we get

$$<n> = N_{3\times\bar{3}}(Y_{3\times\bar{3}}, Q^2_{3\times\bar{3}}) + N_{8\times\bar{8}}(Y_{8\times\bar{8}}, Q^2_{8\times\bar{8}}) \atop {x \to 0 \atop Q^2 \to \infty} \tag{6.15}$$

with the asymptotic prediction from QCD

$$N_{3\times\bar{3}} = \frac{4}{9} N_{8\times\bar{8}}$$

Unifying models based on dual unitarization schemes /171,172/ and on dual topological unitarization /173,174/ arrive at similar conclusions /139/.

Referring to a comparison of hadronic events in e^+e^- annihilations and in current-nucleon reactions, the predictions of (6.14,15) are obvious:

- same height of the rapidity plateau, as measured at $Y_{CMS} = 0$ (e^+e^-) and at $Y_{BREIT} = 0$ (1N), respectively;
- for valence quark scattering, the total multiplicities observed in the two reactions should be identical for same W, up to a small additive constant from a possible difference in quark and diquark fragmentation;
- in 1N reactions the total multiplicity at fixed W increases with decreasing x, because then the relative contribution of the color octet part grows.

Figure 6.19 shows dσ/dy from νp reactions /164/ for $8 \leq W \leq 16$ GeV, compared to e^+e^- annihilations at $W = \sqrt{s} = 13$ GeV /24/. The height of the charged-particle plateau, (dn/dy) ≅ 2, is consistent. In Fig.6.20 the total multiplicity measured in νp reactions /160,164/ is compared with data from e^+e^- annihilations (see Chap.2). Again the values agree surprisingly well, except at very low energies where the phase space available to fragments is significantly reduced in νp interactions due to the existence of a final state nucleon.

Figure 6.21 demonstrates the dependence of $<n_{ch}>$ on Q^2 (or, equivalently, x) at fixed W for νp interactions. In contrast to the expectation based on (6.14,15), $<n_{ch}>$ shows no significant dependence on Q^2. This feature is in basic agreement with other experiments studying deep inelastic lepton-nucleon interactions /175-177/.

When interpreting Figs.6.19-21 in terms of jet universality, one should note that at these energies the multiplicity is far from being dominated by the plateau region. Below W = 5 GeV, the multiplicity is more or less fixed once the transverse momentum smearing of the produced particles is given and is nearly independent on the choice of matrix elements. The invariance of $<n_{ch}>$ with Q^2 will be partially due to this fact. It may further indicate that at these Q^2 either reactions with a coherent spectator system dominate and no octets are separated or that the typical multiplicities of triplet and octet plateaus are similar.

Clearly, the present data are not sufficient to decide whether jet universality holds exactly; nevertheless, since the data are consistent with universality, we shall keep it as a working hypothesis.

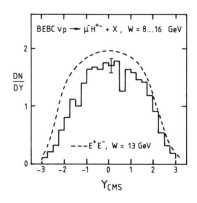

Fig.6.19

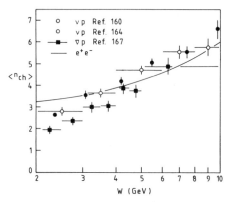

Fig.6.20

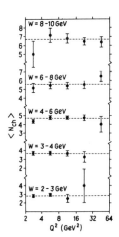

Fig.6.21

Fig.6.19. Rapidity distributions in the cms of secondaries in νp reactions at W = 8-16 GeV /164/, and in e^+e^- annihilations at $\sqrt{s}$ = 13 GeV /24/

Fig.6.20. Mean charged hadronic multiplicity measured in νp reactions and in e^+e^- annihilations

Fig.6.21. Mean number of charged hadrons produced in νN interactions as a function of Q^2 for fixed W /160/

6.4 Spectator Fragmentation

Present data seem to be consistent with the hypothesis of approximate factorization and jet universality. In a certain sense lepton-nucleon reactions therefore don't give qualitatively new information on quark jets as compared to e^+e^- annihilations.

Lepton-nucleon reactions, however, offer the unique possibility to study the fragmentation of compound quark states, such as diquarks. It will be shown that the measurement of their fragmentation functions reveals information on the wave function and dynamics of multiparton states. Furthermore, once the fragmentation properties of multiquark systems are known, e.g., from νN reactions, the mechanisms of other deep inelastic phenomena like muon pair production in hadron-hadron interactions can be reconstructed by studying the quark contents of the spectators left over after the hard process.

A spectator system can be described by its flavor content, its color state, its mass, and by the degree of coherence of its wave function. Table 6.2 gives a list of possible spectators in charged current neutrino-nucleon interactions.

Table 6.2. Spectator systems in ν, $\bar{\nu}N$ interactions. Coherent subsystems of the spectator are put in brackets. Only u and d sea quarks are taken into account

Reaction	Spectator Flavor	Color	Mass
ν-valence quark (Fig.6.1)	(uu)	$\bar{3}$	$0(m_n)$
$\bar{\nu}$-valence quark (Fig.6.1)	(ud)	$\bar{3}$	$0(m_n)$
ν-primordial sea (Fig.6.3a)	(uud$\bar{d}$) or (uudu)	$\bar{3}$ 3	$m_n(1/x-1)$
$\bar{\nu}$-primordial sea (Fig.6.3a)	(uud$\bar{u}$) or (uudd)	$\bar{3}$ 3	$m_n(1/x-1)$
ν-pointlike sea (Fig.6.3a)	(uud)$\bar{d}$ or (uud)u	(8) $\bar{3}$ (8) 3	$m_n(1/x-1)$
$\bar{\nu}$-pointlike sea (Fig.6.3b)	(uud)$\bar{u}$ or (uud)d	(8) $\bar{3}$ (8) 3	$m_n(1/x-1)$

The spectator is thus a much more complex object than a single active quark. To reduce the number of variables, we shall assume the following simplifications based on our knowledge on quark jets.

First, the studies of QCD jets presented in Chap.5 have shown that the fragmentation function of a system consisting of incoherent components can be represented as the incoherent sum of their fragmentation functions, at least as far as fast ($x \gg 0$) fragments are concerned.

Second, we shall assume that factorization holds here as well. That means the distribution of fast fragments depends only on the flavor contents of the spectator, whereas its color contents influences the plateau region.

We are now ready to define the spectator fragmentation function $D_s^h(z)$ which refers to the decay of a coherent parton system. We use a reference frame where the initial nucleon has a large momentum p_0. Since the maximum momentum available to fragments is $p_{\shortparallel MAX} \cong p_0(1-x)$, the natural scaling variable is $z = p_{\shortparallel}/[p_0(1-x)]$. In the cms, we have $z \cong -x_F$, with x_F defined in the usual way as $2p_{\shortparallel}/W$.

Do we expect scaling to hold for spectator fragmentation? The picture of asymptotic freedom, together with the diagrams of Fig.6.1 suggest that the spectator completely ignores the hard process. The energy used to build up the preconfinement plateau is radiated by the active quark; the color charge of the spectator participates only in the final confining step at fixed Q_0^2. Does this imply that the spectator fragmentation function is independent of Q^2?

In fact, a Q^2 dependence of spectator fragmentation arises in a rather indirect, process dependent way as can be seen from Fig.6.6b. The incoming proton dissociates into the active quark at x' and the spectator at (1-x'). The active quark radiates an energy fraction (x'-x) and is then hit by the hard probe. Therefore, the maximum momentum fraction of spectator fragments is (1-x') and not (1-x) as naively expected. Since the amount of radiation (x'-x) increases with increasing Q^2, the spectator fragmentation function D(z) shrinks with increasing Q^2 and the particle density at low z increases. Qualitatively these effects of scale breaking equal those observed in quark fragmentation; quantitatively they differ and depend on x as well as on Q^2. This difference of the perturbative jet structures in e^+e^- and in lN reactions has been emphasized by various authors /181-184/.

In reality, however, these differences are small compared to nonperturbative effects, at least at present values of Q^2 and W. This is evident from Fig.6.22 where the distribution of sphericity is compared for e^+e^- events and for lN reactions in two bins of W. The distributions agree within the error bars. Therefore, we feel that in the following discussion scale breaking effects can be neglected without introducing much bias.

Let us now consider the fragmentation of the coherent part of the spectator. Because its decay is governed by soft, collective processes, one may ask if the perturbative quark-parton model or, more generally, the concept of individual quarks may be applied. Consider the decay of the spectator (mass m_s) into a parton b and a core of mass m_x. Let z be the fractional momentum of b and $p_\perp$ its transverse momentum. From simple kinematics we get

$$p_b^2 = -\frac{p_\perp^2 + zm_x^2}{(1-z)} + zm_s^2 \qquad (6.16)$$

If the core z contains valence constituents of the spectator, its mass should be of the order O(GeV) /189-191/. That means, however, that for z → 1 the active parton is far off shell. A fragmentation which contains this parton thus actually tests the parton structure at large Q^2 and the application of the perturbative parton concept appears to be justified /192/. To conclude: although spectator fragmentation is basically a soft process, the quantum numbers of fast fragments should reflect the constituent structure of the spectator.

Unfortunately, experimental data at sufficiently high Q^2 and W suited to test these ideas are rare. Figure 6.23 shows the x_F distribution of positive and negative fragments observed in νp reactions in the 15 ft bubble chamber /160/ for events with W > 4 GeV and x > 0.05, and with typical Q^2 of the order of 2-10 GeV^2. Similar results have been reported by BEBC /164/. In Fig.6.23 target fragments populate the region $x_F \rightarrow -1$. For comparison, the distributions observed in the proton fragmentation region in inelastic $\pi^+ p$ reactions at $\sqrt{s}$ = 5.6 GeV /185/ are included as dotted lines.

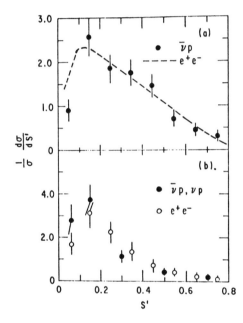

Fig.6.22. (a) Sphericity distribution for $\bar{\nu}p$ reactions with W = 6.6 GeV and for e^+e^- annihilation at W = 7 GeV; (b) sphericity distribution for νp plus $\bar{\nu}p$ reactions at W = 10.5 GeV and for e^+e^- annihilation at W = 13 GeV. From 15' $\bar{\nu}H_2$ /167/

The similarity of the pion spectra obtained in the two reactions is striking. Furthermore, in each case proton production dominates for $x_F \rightarrow -1$. The absolute proton yield, however, is a factor of 5 less in the deep inelastic reaction for $x_F \rightarrow -1$. A dominant production of neutral baryons can be excluded. Baryon number conservation requires the proton distribution to peak at lower x_F, in contrast to the $\pi^+ p$ events.

The full lines in Fig.6.23 are the result of a phase space model (UJM). The model uses constant matrix elements to describe pion production. The matrix element for proton production has been adjusted to fit the experimental proton spectrum.

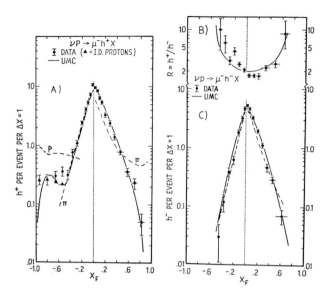

Fig.6.23. (a) Density of positive particles vs Feynman x_F in the hadronic rest frame for νp interactions with $x > 0.05$, and $W > 4$ GeV. From 15' νp /160/. Full lines refer to a longitudinal phase space model. Dotted lines show the hadron distribution in $\pi^+ p$ interactions at $\sqrt{s} = 5.6$ GeV /185/. (b) Ratio of positive to negative hadrons vs x_F. (c) Density of negative hadrons vs x_F

In addition, experimental values for the multiplicities have been used as an input. The agreement between data and the simple model is surprisingly good, demonstrating once more that at those energies, jet physics is still governed mainly by phase space effects.

Nevertheless, the following facts have to be explained by fragmentation models.

- The spectrum of positive pions agrees with that obtained from normal $\pi^+ p$ or pp interactions at similar energies. The ratio of positive to negative pions increases for fast pions. The increase is slightly stronger than observed in normal interactions. However, most of this second effect, if not all, may be accounted for by charge conservation.
- The production of fast protons is suppressed compared to normal inelastic interactions.

Various models have been proposed to describe the fragmentation of multiquark systems, such as generalizations of the FF model /186/, multicomponent fusion models /187/, etc. We shall discuss two models which rely on a minimum number of additional assumptions, are most commonly accepted and well documented, and have proven useful for the description of normal inelastic reactions: the quark recombination model /14/ and the dimensional counting rules /188,193/.

6.5 The Quark Recombination Model (QRM)

The quark recombination model is based on a recent observation by OCHS /13/ that the x_F distributions of fast mesons produced in proton-proton collisions closely resemble the x distributions of quarks known from deep inelastic lepton-nucleon scattering. If these mesons were produced from fast quarks by fragmentation, as in e^+e^- annihilations, the meson spectra would fall much more steeply in x_F than is observed, since one would have to convolute the quark fragmentation function over the probability distribution for quarks in a proton.

This observation led GOLDBERG /194/ and later DAS and HWA /14/ to propose that fast mesons are produced in hadronic reactions by the recombination of valence and sea quarks, at x_q and $x_{\bar{q}}$, respectively, into a meson $q\bar{q}$ at $x_M = x_q + x_{\bar{q}}$. Thus the production of fast mesons measures the combined probability $G(x_q, x_{\bar{q}})/(x_q x_{\bar{q}})$ of finding two quarks in the wave function of the proton:

$$\frac{E}{\sigma}\frac{d^3\sigma_M}{dp^3} = \int \frac{dx_q}{x_q}\frac{dx_{\bar{q}}}{x_{\bar{q}}} G(x_q, x_{\bar{q}}) R(x_M, x_q, x_{\bar{q}}) \qquad (6.17)$$

where σ is the total nondiffractive cross section and R is a recombination function which parametrizes the fusion process. R can be decomposed into a presumably scale invariant two body piece and a many body piece

$$R(x_M, x_q, x_{\bar{q}}) = R_2\left[\frac{x_q}{x_M}, \frac{x_{\bar{q}}}{x_M}\right]\delta(x_M - x_q - x_{\bar{q}}) + R'(x_M, x_q, x_{\bar{q}}) \qquad (6.18)$$

Because of the difficulty of many body recombination, R' is assumed to be negligible except for x_M very close to 1 where it gives rise to Regge behavior. This is visualized by a "competition" argument /195/: fast hadrons are produced after the rapidity plateau of $x \simeq 0$ has been built up /80/. At that time, however, most of the hard gluons which could participate in a multibody recombination will already be "used" by other fragments, i.e., will have turned into quark pairs feeding the particle production at low x.

$R_2(x_q, x_{\bar{q}})$ can be determined from plausibility arguments. One expects recombination to be of short range in rapidity; thus, R_2 will be zero for $|Y_q - Y_{\bar{q}}| \gg 1$. In addition it seems natural to assume that the probability of two quarks at $x_q, x_{\bar{q}}$ to recombine into a meson at x_M is proportional to the probability of finding the two valence quarks of a meson at x_M at $x_{\bar{q}}$ and x_q. That means

$$R_2(\xi_q, \xi_{\bar{q}}) = G_\pi(\xi_q, \xi_{\bar{q}})/(\xi_q \xi_{\bar{q}}) \qquad (6.19)$$

with $\xi_q = x_q/x_M$ and G being the two-valence-quark structure function of the pion. Equation (6.19) can be fulfilled by choosing

$$R_2(\xi_q,\xi_{\bar{q}}) = \alpha \xi_q \xi_{\bar{q}} \qquad (6.20)$$

Equation (6.20) yields a single-quark structure function of the pion

$$G_\pi(x_q) \underset{x_q \to 1}{\sim} (1-x_q) \qquad (6.21)$$

in approximate agreement with experiments /196/. Probability conservation requires $\alpha \leq 4$ /197/ in (6.20).

In practice, σ_M is nearly independent of the specific choice of R_2, once the condition of short range recombination is obeyed /198/.

So far the discussion has been quite general. The QRM seems to be a sort of different phrasing of the fragmentation model discussed in Chap.4. The distinction is that here the attention is centered on the recombination of quarks whose distributions are given a priori, whereas the other model mainly deals with the creation of the quark cloud and treats recombination into mesons as a secondary process.

The new and basic assumption of the QRM concerns the quark distribution functions G(x). Since the production of fast fragments involves large momentum transfers, the G(x) are assumed to be identical to the structure functions measured in deep inelastic interactions. In contrast to the quark fragmentation models, the fast active quark doesn't spend energy to build up a parton sea at low x, since this sea is already "present" in the incident nucleon and its effect is contained in the structure functions.

In the case of proton-proton interactions where sea-antiquarks are concentrated at low x, the two-quark distribution function can be approximated as /197/

$$G(x_q,x_{\bar{q}}) \simeq \lim_{\varepsilon \to 0} G_q(x_q) \bar{x}_{\bar{q}} \delta(x_{\bar{q}} - \varepsilon) \qquad (6.22)$$

where $\bar{x}_{\bar{q}}$ is the average momentum fraction carried by the quark species $\bar{q}$. Equations (6.17,22) explain Ochs observation, they yield for fast mesons

$$\frac{1}{\sigma}\left[x_M \frac{d\sigma_M}{dx_M}\right] \simeq \bar{x}_{\bar{q}} \alpha G_q(x_M) \qquad (6.23)$$

Using the FF parametrization of parton structure functions /199/ the QRM describes extremely well the cross sections for $\pi^{\pm}$ and K^+ production by proton beams for $0.5 \leq x_M \leq 0.9$ /14,197/.

With less theoretical justifaction, the QRM can be extended to lower values of x_M, including the production of K^- in proton-proton interactions via recombination of two sea quarks /201/. At low x_M, however, difficulties arise due to hadron production by resonance decays. At high x, $x \gtrsim 0.4$, resonance contributions to the inclusive spectra were shown to be negligible both experimentally /202/ and theoretically /203/.

Though the QRM gives a good description of the shape of inclusive particle spectra, the absolute magnitudes of the cross sections turn out to be considerably smaller than the data, when calculated for the standard amount of sea quarks. To fit the data, the momentum fraction carried by the quark sea of the nucleon has to be 20 to 50% /14,197,200/. Taking into account that the quark-antiquark system has to be a spin-0 color singlet would increase the required number of sea quarks by another order of magnitude /195/.

The enhancement of the sea is usually explained by the following mechanism /195, 200/. In an undisturbed hadron, sea quarks and gluons form an equilibrium state. During the interaction, this equilibrium is disturbed since quark-antiquark pairs "condense" into mesons. Consequently, new quark pairs are created and the process repeats until all the gluons are used. Therefore, the momentum fraction carried by the effective sea should be of the order of 50% or somewhat less, depending on the time scales governing the quark-gluon equilibrium and the recombination mechanism. Finally, it is not unnatural to assume that the spin and color states of the final mesons will be adjusted by emission and absorption of soft gluons without modifying the z distribution of the mesons.

The ideas nicely dovetail with the probabilistic quark model approach to proton fragmentation by POKORSKI and Van HOVE /204-206/. They assume that in hadronic collisions the gluon clouds of the hadrons interact and generate the rapidity plateau, whereas the incident valence quarks fly through without change of momentum and recombine to leading baryons and fast mesons.

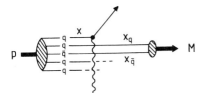

Fig.6.24. Quark recombination in reactions initiated by pointlike probes measures the three quark distribution in the nucleon

The QRM has been applied to deep inelastic reactions by De GRAND et al. /201,207/. The situation is only slightly more complicated in this case; here one measures the combined probability for seeing three quarks (Fig.6.24) in the wave function of the proton.

For instance, the meson spectrum in the target fragmentation region in deep inelastic electron-hadron scattering is

$$\frac{1}{\sigma(x)} \frac{d\sigma(x,z)}{dz} = \frac{\sum_i q_i^2 \int \frac{dx_q}{x_q} \frac{dx_{\bar{q}}}{x_{\bar{q}}} G(x,x_q,x_{\bar{q}}) R_2\left[\frac{x_q}{x_M},\frac{x_{\bar{q}}}{x_M}\right] \delta[x_M-(1-x)z]}{\sum_i q_i^2 G(x)} \quad (6.24)$$

where q_i is the charge of the ith quark and z is the fraction of the recoiling core's momentum (in an infinite momentum frame) carried off by the meson. The cross section is normalized by the total cross section at fixed Bjorken x, i.e., it describes the number of mesons per event. Similar expressions hold for neutrino-nucleon scattering.

Now to arrive at quantitative predictions, one has to make models for the multi-quark distributions in the target.

The simplest model of this type is the KUTI-WEISSKOPF model /208/ which is simply the uncorrelated jet model (Chap.3) applied to partons in a nucleon. In analogy to (3.1) the "transition probability" for a proton into a Fock state containing three valence quarks and N sea partons is given by

$$\Gamma_N \sim f_u(x_u) f_u(x_u') f_d(x_d) \left[\frac{\kappa^N}{N!} \prod_{i=1}^{N} f_s(x_i)\right] \delta(1-\Sigma x) \quad (6.25)$$

Equation (6.25) is written for one type of sea partons, its generalization to n quark flavors and gluons is obvious. The coupling constant κ determines the mean number of sea partons.

The matrix elements f are chosen such that the inclusive quark distributions, obtained from (6.25) by integrating over the N+2 unseen partons and by summing over the N = 0 to N = ∞ Fock states, reproduce the experimental distributions. The coupling constant κ determines the mean fraction of momentum carried by soft sea partons. One obtains approximately /198/

$$G_{VAL.\ QUARK}(x) \underset{x \to 1}{\sim} (1-x)^\kappa f_{VAL.\ QUARK}(x)$$

Parametrizations for the matrix elements f are given, e.g., /198,201/.

Let us now specify the predictions of the QRM for νp scattering (Fig.6.23). The basic features can be derived without explicit calculation. In Kuti-Weisskopf models the two quark wave functions factorize approximately,

$$G(x_1, x_2) \sim G(x_1) G\left[\frac{x_2}{1-x_1}\right] \tag{6.26}$$

Taking x_1 as the x of the quark being scattered, $x_2/(1-x_1)$ is identical to the scaling variable z describing the fragmentation of the spectator. That means the distribution of a valence quark of the spectator is identical to its distribution in the initial proton, if written in the correct scaling variable.

For νp interactions at very low x, the current will mostly scatter off a sea quark, and the spectator contains all three initial valence quarks. As a result of (6.26) the distribution of fast mesonic fragments should be identical to that observed in normal inelastic interactions. Furthermore, the distribution of π^+ and π^- will be similar, except for a normalization factor of 2 since there are two u-quarks, and a factor (1-x) reflecting the dominance of u-quarks at large x. As the x of the scattered quark increases, the neutrino will mostly interact with the d-quark. Therefore, the π^+ distribution will remain unaffected, whereas the π^- distribution steepens since a π^- can no longer be produced by recombination of a valence quark with a sea parton.

The pion and kaon spectra resulting from the exact evaluation of (6.24) /201/ are shown in Figs.6.25,26 for charged current νp interactions. The curves represent the lorentz invariant cross section $(2/\pi W) \int dp_\perp^2 \, E d^3\sigma/dp^3$, and the predictions are extended to low x_F using empirical corrections and sea enhancement factors /201/. The normalization has been adjusted to an electroproduction experiment at lower energies /176/. In Fig.6.26 the ratio of π^+ to π^- production is compared for various Bjorken x, nicely demonstrating the transition from sea quark scattering (low x) to valence quark scattering ("high" x). The corresponding spectra for $\bar{\nu} p$ scattering are shown in Figs.6.27,28 including data from a $\bar{\nu}$ scattering experiment at lower energies.

Since in the latter case the spectator always contains both u- and d-quarks, the spectra of π^+ and π^- are similar.

As far as baryon production is concerned, the predictions of the QRM are rather ambiguous since there are various competing production mechanisms. For scattering of a sea quark at low x, the possibilities are depicted in Fig.6.29a-d. The two extrema are the recombination of all valence quarks to a baryon and the recombination of the valence quarks into three mesons with the baryon being formed by the sea quarks left over. The corresponding graphs for spectators containing two valence

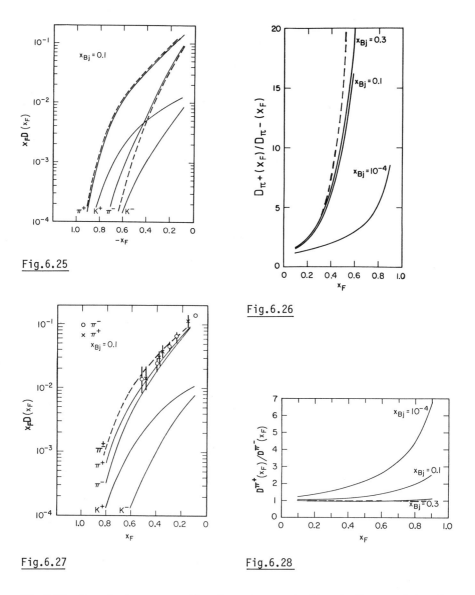

Fig.6.25

Fig.6.26

Fig.6.27

Fig.6.28

Fig.6.25. Invariant cross section for meson production in the spectator fragmentation region in $\nu p \to \mu^- h^\pm x$ reactions as a function of x_F. Full curves are predictions of the QRM /201/, absolutely normalized. Dotted curves are predictions of dimensional counting rules (DCR), arbitrarily normalized

Fig.6.26. Ratio of positive to negative pion density in the spectator fragmentation region in $\nu p \to \mu^- h^\pm x$ as a function of x_F for different values of Bjorken x. Curves as in Fig.6.25

Fig.6.27. As in Fig.6.25, but for $\bar{\nu} p \to \mu^+ h^\pm x$. Data from /167/

Fig.6.28. As in Fig.6.26, but for $\bar{\nu} p \to \mu^+ h^\pm x$

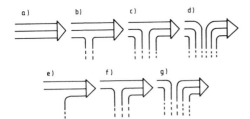

Fig.6.29a-g. Diagrams for baryon production by spectator systems containing all three initial valence quarks (a-d), and two valence quarks (e-g)

quarks of the initial proton are shown in Fig.6.29e-g. Obviously, for $z \to 1$, the graphs with the maximum number of valence quark lines collected in the baryon will dominate.

The relative probabilities can be estimated by noticing that the processes shown in Fig.6.29a are presumably identical to those describing proton fragmentation in normal proton-proton interactions.

An analysis of such interactions in terms of the graphs shown above is reported in /206/. It is stated that the probability T of two incident valence lines emerging in a single hadron is about 0.65. Assuming that T has an universal value for all "soft" hadronic processes /206/, we are now able to predict the shape of proton spectra in the spectator fragmentation region. For sea quark scattering, the terms dominant at large z (Fig.6.29a,b) generate T^2 and $(2T)/2$ protons per event, in the mean. The factor 1/2 in Fig.6.29b arises since roughly half of baryons will be neutrons. For valence quark scattering, Fig.6.29e dominates at high z, yielding an integral number of T/2 protons per event. The QRM can be easily generalized to this type of recombination yielding,

$$\left[\frac{d\sigma_p}{dz}\right]_{z \to 1} \sim G_i(z) \tag{6.27}$$

with $G_i(z)$ being the distribution of the momenta of parton systems containing i valence quarks. The normalization is given by

$$\int E\frac{d^3\sigma_p}{dp^3}\frac{d^3\sigma_p}{E} = T^2; \; T; \; \frac{T}{2} \quad \text{(Fig.6.29a,b,e)} \tag{6.28}$$

We are now able to compare the QRM with the high statistics data from the 15' νp experiment (Fig.6.23). The cut $x > 0.05$ provides a rather clean sample of ν valence quark interactions. Figure 6.30 shows the data compared to absolute QRM predictions. The shape of the pion spectra and the charge ratios are well described. The absolute

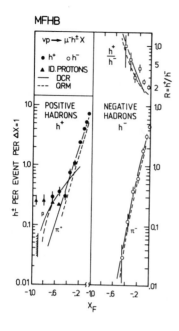

Fig.6.30. Same data as Fig.6.23, compared to model predictions for the spectator fragmentation region $0 \geq x_F \geq -1$. (---) QRM; absolutely normalized /201/; (———) dimensional counting rules. Proton spectra are absolutely normalized; one normalization constant for meson spectra chosen to fit the data

normalization seems to be a bit too low. However, taking into account that the model refers to scaling spectra at very high energies and does not contain non-asymptotic corrections, the agreement is fairly good. The proton distribution seems to drop slightly too fast as $|x_F| \to 1$. However, one should note that for the mean W of this data, the kinematical limit for proton production is $x_F = -0.90$ to -0.95, so at least the data point at highest $-x_F$ appears somewhat questionable. Secondly, the excess at high x may be due to a small remaining fraction of diffractive events. This is supported by the BEBC data (Fig.6.31) which indicates that the proton cross section drops at $x_F \to -1$ as Q^2 increases.

6.6 Dimensional Counting Rules

A second and widely used approach to describe the fragmentation of multiparton systems is based on dimensional counting techniques /209-213/. The dimensional counting rules (DCR) developed by GUNION /214/, BLANKENBECLER and BRODSKY /215/, FARRAR /216/, and BLANKENBECLER et al. /217/ refer to the inclusive spectra of the fragment a in the fragmentation process

$$A \to a + X \qquad (6.29)$$

BEBC $\nu p \rightarrow \mu^- h^\pm X$

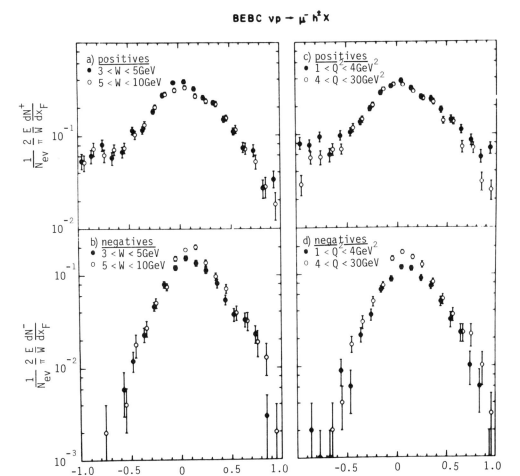

Fig.6.31a-d. Normalized lorentz invariant cross section for positive (a,c) and negative (b,d) particle production in νp charged current interactions from BEBC νH_2 /164/ for 2 intervals of W and of Q^2

As usual, the process is described in a frame where the momenta p_A and p_a are large. DRC predict

$$\frac{1}{\sigma} \int E \frac{d^3\sigma_a}{dp_\perp^2} dp_\perp^2 \sim z \frac{1}{\sigma} \frac{d\sigma_a}{dz} = z D_A^a(z) \underset{z \to 1}{\sim} (1-z)^m \tag{6.30}$$

with $z = p_a/p_A$ and $m = 2n_x-1$, where n_x is the minimum number of quarks left behind in X.

Equation (6.30) can be derived using the Bethe-Salpeter equation to describe an n-quark bound state /201,212,215/. If such a state decays into a particle a and n_x residual quarks, the mean momenta of the residual quarks go to zero as $(1-z)$ for $z \to 1$. This gives rise to a phase spacelike suppression factor $(1-z)^{n_x}$ in the transition amplitude, or $(1-z)^{2n_x}$ in the decay probability. In the overall expression, one factor $(1-z)$ is cancelled by an energy denominator referring to the core energy. The spectrum of fragments a is then given by

$$\frac{1}{\sigma} z \frac{d\sigma_a}{dz} = \sum_{i=n_x}^{\infty} f_i (1-z)^{2i-1} \qquad (6.31)$$

with f_i being the probability that i quarks are left behind.

Equation (6.31) holds only for 1-z small, e.g., $z \gtrsim 0.5$ /201,216/. In this limit, σ_a will be dominated by the leading term shown in (6.30).

There are basic differences between the DCR ansatz and the QRM /201/. In the counting rule scheme, no distinction is made between sea and valence quarks of the incident particles; both are treated on equal footing and with identical matrix elements. The different shape of valence quark structure functions arises from the summation over the different Fock states. In each of these states valence quarks are present, whereas the mean number of sea quarks equivalent to the relative importance of higher Fock states is small. Consequently, for semi-inclusive deep inelastic scattering we assume the nucleon breaks up into a quark c (which is struck by the current) and a core A of n_A quarks. This core then breaks up into a meson a and a recoiling core of n_x quarks. Thus, we get in analogy to (6.31)

$$\frac{1}{\sigma(x)} z \frac{d\sigma_a(x,z)}{dz} \underset{z \to 1}{\sim} \frac{\sum_{n_A} f_{n_A} (1-x)^{2n_A-1} (1-z)^{2n_x-1}}{\sum_{n_A} f_{n_A} (1-x)^{2n_A-1}} \qquad (6.32)$$

This implies, however, that between the hard scattering and the fragmentation the quark core A again reaches an equilibrium state.

Note that in (6.31,32) the power in (1-z) may increase by one unit due to spin effects /212,218/.

Let me demonstrate the application of DCR in a few examples.

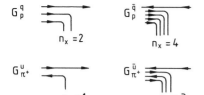

Fig.6.32. Counting rule graphs describing nucleon and meson structure functions. n_x is the number of residual quarks

Figure 6.32 shows the leading graphs contributing to the proton and pion structure functions. For $x \to 1$, we predict

$$G_p^{u,d} \sim (1-x)^3$$
$$G_p^{\bar{u},\bar{d},\ldots} \sim (1-x)^7$$
$$G_{\pi^+}^{u,\bar{d}} \sim (1-x)$$
$$G_{\pi^+}^{\bar{u},d,\ldots} \sim (1-x)^5$$

(6.33)

in reasonable agreement with the experimental values. Figure 6.32 can as well be read as defining quark fragmentation functions (from right to left). One gets

$$G_A^a(x) \underset{x=z \to 1}{\sim} D_a^A(z) \qquad (6.34)$$

as required by rather general correspondent arguments /99,100/.

The discrepancy between the prediction $D_u^{\pi^+} \to (1-z)$ and the experimental results, close to $(1-z)^2$, can be explained by spin effects. The predicted suppression factor for unfavored decays, $(1-z)^4$, however, is in clear disagreement with the observed behavior $(1-z)$. A possible explanation is that DCR does not include effects such as resonance production. Favored production of ρ mesons by an u-quark, followed by a decay into $\pi^+\pi^-$, yields a π^- spectrum suppressed just by $(1-z)$ as compared to the ρ spectrum. These modes have to be taken into account when applying DCR. They are especially important in quark jets where the ratio of vector meson to scalar meson production is of the order of 1 /87/.

Let us now calculate the distribution of spectator fragments in νp and ν̄p charged current interactions with valence quarks. We obtain

$$\nu p: \int dp_\perp^2 E \frac{d^3\sigma}{dp^3} \sim \begin{array}{l} zD^p_{uu}(z) \sim (1-z) \quad \text{for protons} \\ zD^{\pi^+}_{uu}(z) \sim (1-z)^3 \quad \text{for } \pi^+ \\ zD^{\pi^-}_{uu}(z) \sim (1-z)^7 \quad \text{for } \pi^- \end{array}$$

and (6.35)

$$\bar{\nu} p: \int dp_\perp^2 E \frac{d^3\sigma}{dp^3} \sim \begin{array}{l} zD^p_{ud}(z) \sim (1-z) \quad \text{for protons} \\ zD^{\pi^\pm}_{ud}(z) \sim (1-z)^3 \quad \text{for } \pi^\pm \end{array}$$

Figures 6.25-28 show these curves together with the QRM predictions; a comparison with νp data is included in Fig.6.30. In the last case, the proton cross section has been normalized to give 0.5 protons per event. The pion spectra are arbitrarily normalized, however, the same normalization constant is used for both π^+ and π^- spectra (although this is not necessarily required by DCR). Except for the proton spectra at high $|x_F|$, the agreement with data is excellent.

Unfortunately, the fact that both QRM and DCR roughly agree with data does not help to shed light on the underlying fragmentation mechanism. The two models are even contradictory. The QRM assumes that the final mesons reflect the quark distribution immediately after the interaction, whereas the interpretation of the DCR requires the spectator partons to reach a new equilibrium state. We shall return to this point in Chap.7.

Let me now briefly go through a few recent extensions of DCR. The influence of spin effects has been clarified in a recent work by BRODSKY /219/. If the helicities of parent and fragment differ by Δh, an additional factor $(1-z)^{2|\Delta h|}$ arises in $d\sigma/dz$. This correction improves the agreement with experiment for the favored quark fragmentation into mesons.

Furthermore, one has to ask what scale invariant counting rules mean in a world where scaling seems to be violated at least as far as structure functions are concerned. It has been shown by FRAZER and GUNION /220/ that using the DCR approach via the Bethe-Salpeter equation reproduces the Altanelli-Parisi equations. A heuristic picture to this is given in the model of scale breaking by Kogut and Susskind (Chap.5). Increasing Q^2 yields an improved resolution of the system under study so that effectively the system appears to have more constituents, resulting in a steepening of the structure and fragmentation functions

$$zD(z) \cong (1-z)^{2n_{eff}(Q^2)-1} \tag{6.36}$$

Additional violations of scaling, which appear in systems with an even number of constituents and vanish like $1/Q^2$, have been discussed by VAINSHTAIN and ZACHAROV /218/.

Finally, the question of the time scales involved in parton fragmentation has been investigated by GUNION /221/. As previously mentioned, there seems to be a disagreement between the requirement of an intermediate equilibrium state in DCR and the fact that particle production at x close to 1 involves off shell partons (6.16) resulting in short decay times (measured in the rest system of the fast fragment). Since this process will contain large momentum transfers, the quark pairs required for the fragmentation can be created in a pointlike manner during the decay (Fig.6.33a). These quarks do not participate in the momentum sharing among partons belonging to the n-quark Fock state of the parent system, the suppression per quark line left over is smaller than in the original DCR. Using these "pointlike" counting rules one obtains

$$D_A^a(z) \sim (1-z)^{2n_x+n_{pl}-1} \tag{6.37}$$

where n_x is the number of those quarks in X which are completely unaffected by the decay, and n_{pl} is the number of quarks in X which participate in a hard creation process. In the case of νp scattering at not too small x, one obtains now (Fig. 6.33b-d)

$$\begin{aligned} D_{uu}^{\pi^+} &\sim (1-z)^3 \\ D_{uu}^{\pi^-} &\sim (1-z)^5 \\ D_{uu}^{p} &\sim (1-z)^1 \end{aligned} \tag{6.38}$$

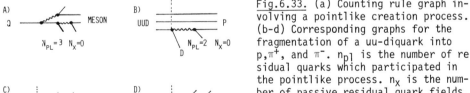
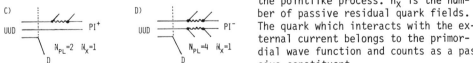

Fig.6.33. (a) Counting rule graph involving a pointlike creation process. (b-d) Corresponding graphs for the fragmentation of a uu-diquark into p, π^+, and π^-. n_{pl} is the number of residual quarks which participated in the pointlike process. n_x is the number of passive residual quark fields. The quark which interacts with the external current belongs to the primordial wave function and counts as a passive constituent

The x dependence in (6.38) has been calculated assuming that the emission of a fast fragment immediately follows the hard scattering, resulting in quark distributions which are identical to those in the incident proton. To derive multiquark distributions, factorization in the sense of (6.26) was used.

Within the limited statistics available and due to the restricted x range for negative fragments, "standard" and "pointlike" counting rules cannot yet be distinguished experimentally.

6.7 The Three Gluon Decay of the T

Up to now we restricted our attention to parton systems where, in a suited reference frame, the momenta of color sources are more or less collinear. The principles of factorization and universality were used to describe the fragmentation of these systems. QRM and DCR provide convenient, if not yet fully understood parametrizations of the fragmentation functions of multiquark systems. There is no ambiguity of choosing the "best" reference frame for the description of the fragmentation, except for longitudinal boosts. The situation changes drastically as soon as three or more color sources are distributed in space.

As an example, we shall discuss the decays of heavy quark-antiquark bound states like the T or the T'. These mesons were discovered /222/ as broad enhancements in the mass spectrum of muon pairs produced in proton-nucleus collisions. The interpretation as bound states of a new quark type has been confirmed by experiments at the e^+e^- storage rings of DORIS /223-226/ and CESR /227,228/. The narrow decay width of the T, $\Gamma \sim 40$ keV /229/, implies that the quarks in the T carry a new conserved flavor, called bottom, and that the T mass is below the threshold for production of naked bottom. Allowed decay modes are annihilations into an odd number of strong, electromagnetic, or weak vector bosons. Relevant decays are the annihilations into a virtual photon, $\Gamma \cong O(\alpha)$, and the annihilation into the simplest color singlet formed of an odd number of gluons, $\Gamma \cong O(\alpha_s^3)$ (Fig.6.34).

Modes with more than three gluons are suppressed since $\alpha_s(M_T^2)/\pi < 1$. In principle the T may also decay into a large number of soft gluons, which correspondingly have

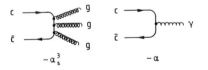

Fig.6.34. Allowed decay modes of the T meson into vector bosons

large couplings. This process is generally assumed to be negligible /230-232/ as those gluons would have a wavelength large compared to the size of the T. Therefore, the T acts effectively as a color singlet and decouples. Since α_S is very large compared to α, the three-gluon decay of the T essentially will saturate the hadronic cross section /233/. The hadronic decays of the T offer the opportunity to study decay properties of noncollinear arrangements of basic color sources; in parallel we may hope to get further insight into the way a gluon fragments.

The matrix element for the decay has been calculated by various authors /234-239/

$$\frac{1}{\sigma} \frac{d^2\sigma}{dx_1 dx_2} = \frac{1}{(\pi^2-9)} \left[\frac{(1-x_3)^2}{x_1 x_2} + \frac{(1-x_2)^2}{x_1 x_3} + \frac{(1-x_1)^2}{x_2 x_3} \right] \quad (6.39)$$

where x_i is the fractional momentum of the ith gluon, $x_i = 2|p_i|/M_T$. The inclusive x distribution of gluons is shown in Fig.6.35; their angular distribution is illustrated by the Dalitz plot (Fig.6.36). In general, the gluons will be noncollinear. Note, however, that the matrix element (6.39) refers to massless gluons, whereas in the preconfinement picture they are expected to be off shell, $p^2 \lesssim O(M_T^2)$.

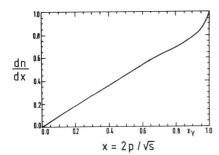

Fig.6.35. The inclusive momentum distribution of on shell gluons from the T decay, according to (6.39)

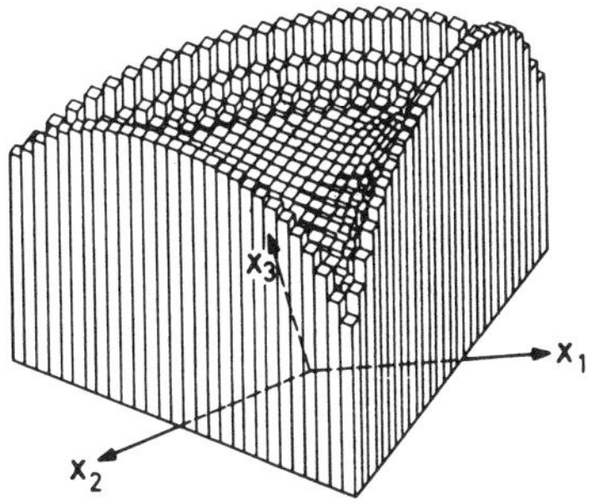

Fig.6.36. Dalitz plot for the decay of the T into three on shell gluons, according to (6.39)

How do these gluons transform into hadrons? As a first step, use the picture that colored objects at large distances from each other build up tubes of color fields. In the case of two partons the tube is obviously along the line connecting them, and the jet axis will coincide with this line. For three partons in a noncollinear configuration, the color field tubes emanating from the partons have to join somehow in order to form a color singlet. It seems natural to assume that the system is in the energetically most favorable state with a minimum of energy stored in the color field.

The actual configuration depends on the properties of the color strings, especially on their tension which is defined by the energy per unit length at rest. In the MIT bag model the tension of an octet or gluon tube is $r = 3/2$ times bigger than that of a triplet tube /240/, lattice gauge theories predict $r > 2$ /241/.

In the latter case, it is more favorable for an octet or gluon to split into triplet tubes.

For the T decay, possible configurations of flux tubes are shown in Fig.6.37. In case of $r < 2$, the octet tubes join in the middle, and a "color center frame" can be defined as that frame where the tensions of the three strings are in equilibrium /242/. In the center of the color frame, the gluons are emitted with angles of 60^0 relative to each other /242/.

For $r \geq 2$ the octet split up and the gluons are connected by three triplet tubes /241,242/.

Corresponding diagrams are obtained for $q\bar{q}g$ states resulting from e^+e^- annihilations (Fig.6.38). In the limit $r \geq 2$, the gluon may be identified as a "kink" in the triplet string joining the quarks. This interpretation of a gluon has been discussed extensively in /243/. It offers a bridge to topological models /174/ where gluons do not appear explicitly. From this point of view, one may even think of the three gluon intermediate state in the T decay as a closed triplet tube with three kinks!

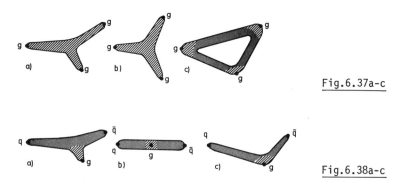

Fig.6.37a-c

Fig.6.38a-c

<u>Fig.6.37a-c.</u> Configuration of color flux strings in the T decay (a) $r \leq 2$, cms frame (b) $r \leq 2$, color center frame (c) $r \geq 2$, cms frame. Shaded and dotted regions are octet and triplet tubes, respectively

<u>Fig.6.38a-c.</u> As in Fig.6.37, but for $q\bar{q}g$ jets

In analogy to the Schwinger model (Sect.4.2), the hadron distribution may be calculated as the four-dimensional Fourier transform of the color field strength (4.10). Equivalently, we may say that the color tubes decay into hadrons with limited momentum transverse to the color-anticolor axis rather than to the jet axis in the overall cms. In the following, we shall refer to three models for the T decay:

Model I: Each gluon fragments as a color octet, and the process is described in the overall cms. If a gluon has a momentum p, an octet-antioctet jet is generated with an invariant mass $M_{8 \times \bar{8}} = 2p$, and one half of it is taken as the gluon jet. This straightforward picture has been used by many authors /234,235,237/.

Model II: Assume $r \leq 2$. The octet-octet strings join symmetrically in the center of color frame. Gluon fragmentation is described as above, but in the center of color frame.

Model III: Assume $r \geq 2$. Then the octet strings split into three triplet strings, the fragmentation of each is described in its own center of mass frame. The triplet strings are treated as quark-antiquark jets, the "decay" of the gluon into quarks being described by the usual QCD cross section.

In models II and III, the lorentz transformation to the overall cms will lead to an increase of the mean transverse momentum squared with respect to the new jet axis in the cms and will modify the distribution of energy flux /244/. However, the factorization property (6.6) will still hold for fast particles since the scaling variable transforms as

$$z_{CMS} = z_{COLOR\ CENTER} + O\left(\frac{m_{\perp cms}}{p_{\parallel cms}}\right) + O\left(\frac{m_\perp}{p_\parallel}\right) \tag{6.40}$$

where $m_\perp$ and $p_\parallel$ refer to quantities in the color center frame.

A more sensitive quantity is the mean decay multiplicity of the T. For model I we get

$$\langle n \rangle = \frac{1}{\sigma} \int_0^1 dx \left[\frac{d\sigma_g}{dx}\right] \frac{1}{2} n_{gg}(x\sqrt{s}) \tag{6.41}$$

where $n_{gg}(\sqrt{s})$ is the mean multiplicity of a gluon - "anti"gluon jet system of mass $\sqrt{s}$. In model II, the expression simplifies to

$$\langle n \rangle = 3 \cdot \frac{1}{2} n_{gg}(\tfrac{2}{3}\sqrt{s}) \tag{6.42}$$

Finally, for model III we have

$$\langle n \rangle = \frac{1}{\sigma} \int_0^{\sqrt{s}} dm \left(\frac{d\sigma}{dm}\right) n_{q\bar{q}}(m) \qquad (6.43)$$

with $d\sigma/dm$ being the mass distribution of the quark-antiquark subsystems. In the limit of on shell gluons, neglecting transverse momenta in the process $g \to q\bar{q}$, $(d\sigma/dm)$ is independent of the choice of the reference frame.

Defining now $R(p)$ as the ratio of gluon to quark jet multiplicities for a parton momentum p,

$$R(p) = n_{gg}(2p)/n_{q\bar{q}}(2p) \qquad (6.44)$$

we estimate for $\langle n \rangle$ from Υ decays,

Model I $\langle n \rangle = (1.10 \pm .02) R\, n_{q\bar{q}}(M_\Upsilon)$

Model II $\langle n \rangle = (1.20 \pm .05) R\, n_{q\bar{q}}(M_\Upsilon)$

Model III $\langle n \rangle = (1.50 \pm .10) R\, n_{q\bar{q}}(M_\Upsilon)$

R is to be taken at $p \simeq O(M_\Upsilon/3)$.

Using the canonical value from QCD, $R = 9/4$, the models I and II predict a tremendous rise in multiplicity on the Υ resonance, as compared to the two-jet continuum. At modest energies, however, phase space effects will decrease the value of R. This can be easily demonstrated. The mean number of particles produced is given by $\langle n \rangle = \sqrt{s}/\langle E_h \rangle$. Two terms contribute to the mean hadron energy $\langle E_h \rangle$: the momentum distribution parallel to the jet axis, depending on the scale invariant matrix element; and the transverse momentum smearing. At moderate energies, the latter is the

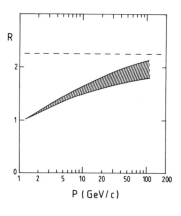

Fig.6.39. Ratio of gluon to quark jet multiplicities as a function of the partons momentum p, from longitudinal phase space models, adjusted to give $R(\infty) = 9/4$

dominant term, and <n> becomes independent on the matrix element. Figure 6.39 shows R as a function of p as obtained from phase space models described in Chap.3. The matrix elements were adjusted to yield $R(\infty) = 9/4$. The approach to asymptotic values is extremely slow. In our range of interest we obtain $R \cong 1.2$, which is in agreement with results obtained from QCD models where the gluon decays via successive branching (Fig.5.11). This leads to the expectation for the relative change in multiplicity on the Υ, compared to the continuum:

$$\Delta n/n_{q\bar{q}} \cong \begin{array}{l} 30\% \text{ model I} \\ 45\% \text{ model II} \\ 50\% \text{ model III} \end{array} \qquad (6.45)$$

Consequently, one expects the momentum distribution to steepen slightly.

Let us now consider the experimental data. There the Υ resonance sits on a continuous background from $q\bar{q}$ jets. Since the events cannot be separated on an event to event basis, all experimental quantities refer to a mixture of $q\bar{q}$ and three gluon jets. It is, however, possible to subtract the contribution of $q\bar{q}$ jets from the continuum and from electromagnetic decays of the Υ itself on a statistical basis. In the following, we shall refer to uncorrected values as "on the Υ" and corrected ones as "Υ direct".

Figure 6.40 shows the mean observed multiplicity in the DASP detector as a function of energy /225/. A slight increase in $<n_{obs}>$ is seen in the Υ region. The present data on multiplicities is summarized in Table 6.3.

Table 6.3. Increase of multiplicity for direct Υ decays compared to two jet events

Group	Ref.	$\delta(\%)$	Remarks
DASP II	/225/	12 ± 3	not corrected for acceptance effects, however, corrections should be small (±1)
DHHM	/245/	13 ± 3	
PLUTO	/246/	27 ± 8	corrected values

These values are incompatible with the predictions of model III, they can be accommodated by models I and II only by choosing $R \cong 1.0$.

Figure 4.42 shows the invariant cross section for the reactions

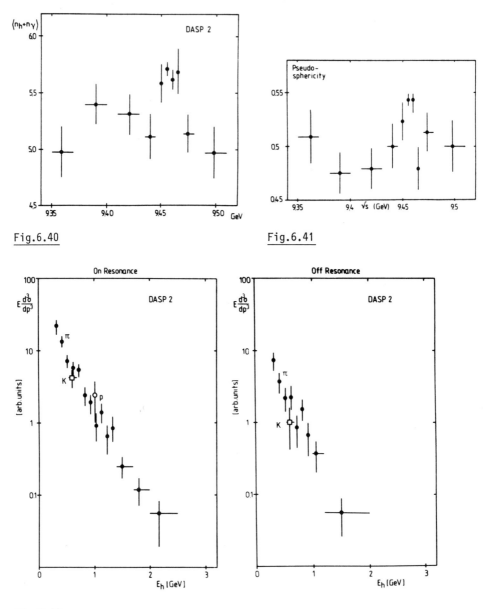

Fig.6.40

Fig.6.41

Fig.6.42

Fig.6.40. Average visible multiplicity in the DASP inner detector as a function of $\sqrt{s}$ /225/

Fig.6.41. Mean pseudosphericity as a function of $\sqrt{s}$. From DASP /225/

Fig.6.42. Invariant production cross section for the inclusive reactions $e^+e^- \to \pi^\pm x$, $K^\pm x$, $p^\pm x$ on and off the $\Upsilon(9.46)$ resonance. From DASP /225/

$$e^+e^- \to \pi^\pm x, K^\pm x, p^\pm x$$

as a function of the particle energy on and off the resonance as measured by the DASP II group /225/. Above the momentum cutoff of 200 MeV/c, the pions show a purely exponential spectrum, $Ed^3\sigma/dp^3 \sim \exp(-E/E_0)$, with

E_0 (on resonance) = (260 ± 25) MeV
E_0 (off resonance) = (240 ± 25) MeV

Within the statistical accuracy, there is no difference between the values on and off the resonance.

More precise data are available from the PLUTO detector. Figure 6.43 shows the distribution of momenta of secondaries for the Υ resonance and the continuum /248/.

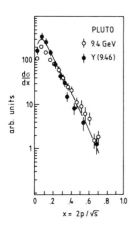

Fig.6.43. Charged particle momentum spectra for the Υ (9.46) and the nearby continuum; the relative normalization of the two data sets is chosen to fit the corrected mean multiplicities (from PLUTO /248/). The full curve is the result of a MC calculation assuming $D_g(z) = D_q(z)$

Direct Υ decays yield a smaller number of particles at high x. Assuming that factorization holds, a model independent guess for particle spectra is obtained by convoluting $d\sigma/dx_g$ over the gluon fragmentation function $D_g^h(z)$. Justified by the small value of (R-1), we chose as a first approximation $D_g^h(z) = D_q^h(z)$. The resulting spectrum is included in Fig.6.43, it fits the data quite well.

The above data do not allow one to decide whether the Υ really decays predominantly into three jets. Instead, one has to study topological quantities like sphericity or thrust. At this place it should be pointed out that Υ decays will not show a pronounced three jet structure, since the mean energy per jet, $\langle E \rangle \cong 3$ GeV, is just at the threshold where it is possible to recognize a jet structure on a statistical basis averaged over many events. The following results rely on a comparison with Monte Carlo models, taking into account the detector acceptance, the reliability of the track recognition, and radiative corrections.

Let me first discuss results from DASP II /225,247/. Since the DASP detector measures directions of particles, but no momenta, four "pseudo" topological variables have been used:

$$\text{Pseudosphericity} = (3/2)\min \langle \sin^2\theta \rangle$$
$$\text{Pseudospherocity} = [(4/\pi)\min \langle |\sin\theta| \rangle]^2$$
$$\text{Pseudothrust} = \max \langle |\cos\theta| \rangle$$
$$\text{Pseudoacoplanarity} = 4(\min \langle |\cos\theta| \rangle)^2$$

where θ is the angle of each track with respect to the preferred axis or, for pseudoacoplanarity, with respect to the preferred plane. $\langle \ \rangle$ means the average over particles of one event. Figure 6.41 shows the mean pseudosphericity as a function of $\sqrt{s}$. Obviously, events from Υ decays are less jetlike. Table 6.4 summarizes the relative changes of these variables on and off the Υ resonance compared to a Monte Carlo calculation. The model, based on the FF algorithm, assumes gluons are produced according to (6.39) and decay exactly like quark jets.

Table 6.4.

Change on/off resonance	Exp. /247/	Model /247/
⟨Pseudoacoplanarity⟩	18.8 ± 4.0	23.0
⟨Pseudosphericity⟩	9.1 ± 1.8	9.8
⟨Pseudospherocity⟩	9.5 ± 2.0	14.0
⟨Pseudothrust⟩	- 2.8 ± 0.6	- 3.1
⟨Multiplicity⟩	8.4 ± 2.0	7.0

The agreement is fairly good, $\chi^2 = 1.4/\text{NDF}$. Υ decays are definitively different from normal two-jet events, a fit in terms of two-jet models gives $\chi^2 = 22/\text{NDF}$. The DASP II group pointed out that if the Υ decays into three gluons, the decay properties of these gluons are identical to those of quarks. A change of the fragmentation function from $D_g(z) = D_q(z)$ to $D_g(z) \sim (1-z)D_q(z)$, as predicted by asymptotic QCD, completely spoils the agreement. Similarly, a significant change in the mean transverse momentum with respect to the jet axis can be excluded.

Recent and very detailed studies of the hadron distribution from T decays by the PLUTO /248/ and DHHM /246/ groups strongly support this picture of a three-jet decay of the T. This is demonstrated by a three-jet analysis using the triplicity method /249/ as performed by the PLUTO group. The triplicity method assigns three jet axes to each event by grouping the detected particles into three disjunct groups in a way as to maximize the sum of the momenta of the jets. Let $x_1 \geq x_2 \geq x_3$ be the normalized jet momenta and $\theta_1 \leq \theta_2 \leq \theta_3$ the angles between the jets. Figure 6.44 shows the experimental distributions of thrust, x_1, x_3, θ_1, and θ_3 for direct T decays and for two-jet events from the continuum /248/. Included are Monte Carlo model predictions for two-jet, three-jet and phase-space like events. Consider first the two-jet data off resonance. An attempt to assign three jet axes to a two-jet event will lead to one jet axis coinciding with one of the jets, and the two other axes pointed in opposite directions. Consequently, x_1 is peaked close to 1, and θ_3 is close to 180°. The distribution in x_3 should be rather flat, since the momentum of one original jet is shared among two reconstructed jets, and the angle between these jets, θ_1 should be small. This is exactly what is observed. For direct T decays, θ_3 decreases and θ_1 increases, proving that it is reasonable to assign three distinct axes. Data are in good agreement with the three-gluon decay model (which assumes that gluon jets behave like quark jets); a description by pure phase space seems to be ruled out.

It has been pointed out that the spin of the three decay partons can be determined from the angular distribution of the axis of the fastest jet with respect to the beam axis /232/. In the limit $x_1 \rightarrow 1$, the distribution of the angle θ between the event axis and the beam direction is given by

$$d\sigma/d\cos\theta \sim 1 + a \cos^2\theta \tag{6.46}$$

with

$a \cong 1$ for spin 1 partons
$a \cong -1$ for spin 0 partons.

Averaged over all values of x_1, one predicts $a \cong 0.39$ /232/ for spin 1 partons and $a < 0$ for spin 0 partons. The distribution of the sphericity axis of direct T decays is shown in Fig.6.45 as measured by the PLUTO and DHHM detectors. Taking into account that the DHHM detector does not allow a precise measurement of the momenta of fast charged hadrons and thus has the tendency to smear the distribution towards isotropy, the spin 1 assignment seems to be slightly favored.

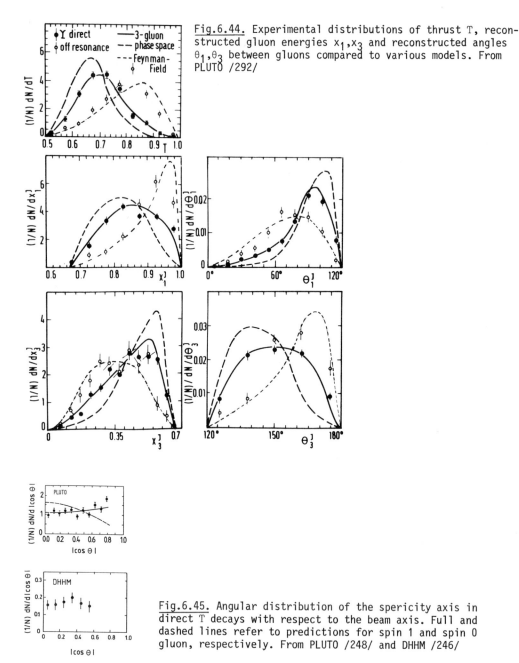

Fig.6.44. Experimental distributions of thrust T, reconstructed gluon energies x_1, x_3 and reconstructed angles θ_1, θ_3 between gluons compared to various models. From PLUTO /292/

Fig.6.45. Angular distribution of the spericity axis in direct Υ decays with respect to the beam axis. Full and dashed lines refer to predictions for spin 1 and spin 0 gluon, respectively. From PLUTO /248/ and DHHM /246/

Let me briefly discuss another prediction from QCD. Since the gluon couples to all quark species with roughly the same strength, in contrast to a virtual photon, the production of strange quarks should be enhanced. However, as pointed out in Chap.5, the effective change will be small. Figure 6.41 demonstrates that in fact no abnormal production of charged kaons is observed on the T. In Fig.6.46 the cross section for neutral kaon production is compared to the total cross section. The relative increase on the T is statistically not very significant, but would be compatible with the expectations.

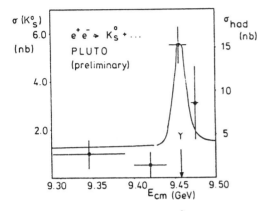

Fig.6.46. Cross section for K^0 production compared to σ_{hadr} as a function of $\sqrt{s}$. From PLUTO /248/

6.8 Summary

The fragmentation of systems containing more than two elementary color sources and color sources composed of two or more quarks, have been studied.

As far as quark jets produced in lepton-nucleon interactions are concerned, the principles of scaling, factorization (or environmental independence), and universality hold approximately and can be used to relate quark jets from different sources and energies. Small violations of scaling and factorization are observed and agree qualitatively with QCD predictions when $O(\alpha_s)$ terms are taken into account. However, regarding the low mean W, it seems more likely that these effects correspond to the "scale breaking" seen in e^+e^- annihilations at very low energies and result from phase space effects due to the nonnegligible particle masses.

The fragmentation of the multiquark spectator systems in lN interactions is shown to be described by quark counting rules or by a quark recombination model.

Let me now summarize what we learned from the study of T decays. The dominance of a three-parton decay mode, probably with spin 1 partons, seems to be established. Although present data have a limited statistical accuracy, there seems to be a disagreement between the expected properties of gluon jets and the observed fragmentation modes which are identical to those of quarks, especially as far as multiplicities are concerned. However, parton fragmentation at these energies is expected to be strongly influenced by phase space effects. Because of these problems, it is hard to draw definite conclusions on the dynamics of systems with noncollinear color sources. Only models where the gluon splits up into two incoherent quark jets seem to be excluded.

Assuming that the decay partons of the T fragment like quarks, data are consistent with factorization.

7. Jets in Hadron-Hadron Interactions with Particles of Large Transverse Momentum

Both in e^+e^- annihilations and in lepton-nucleon scattering jets are produced since a parton interacts with a current of large Q^2. The four-momentum transfer is given by the change in momentum of the lepton and is well defined experimentally (Fig.7.1a). In close analogy, one predicts the interactions of two partons inside two colliding hadrons via the exchange of a vector boson /11/ simply by replacing the lepton at the upper vertex in Fig.7.1a by another quark (Fig.7.1b). The main difference is that the current now may be a gluon as well.

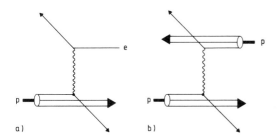

Fig.7.1. Quark diagrams for (a) deep inelastic lepton-nucleon interactions; (b) quark elastic scattering in proton-proton collisions

As a result of the hard scattering, four color sources are distributed in the plane defined by the collision axis of the incoming hadrons and by the three-momentum component of the current. In general, the two active partons will be scattered at large angles in the overall cms, and the spectator systems will move along the directions of the primary hadrons.

When the color sources start to separate in space, flux tubes are built up which in turn "decay" and form jets. From our present knowledge, the configuration of these flux tubes cannot be calculated. It depends on the kinematics of the interaction and on the specific tensions of color octet and triplet strings. In addition, one may have the possibility to group the color sources into two colorless clusters /242/. It seems natural to assume that the flux tubes are formed in a way as to minimize the energy stored in the field /241/. Figure 7.2 shows examples for the case of quark-quark scattering. Depending on the ratio r of the tensions of octet and triplet flux tubes, different configurations are favored. However, since the final distribution of fast fragments depends only on the state of the color field in the proximity of the color charge and on the boost connecting the rest frame of the color charge and the cms, fractorization is still expected to hold. The change of the field for $r \lesssim 2$ concerns only those slow particles in regions of phase space where the four jets join.

The resulting configuration consists of two jets containing particles with large transverse momentum with respect to the collision axis and of two jets of low $p_\perp$ particles (Fig.7.2d). At sufficiently large $p_\perp$, such processes should dominate the particle production in hadronic interactions /11/.

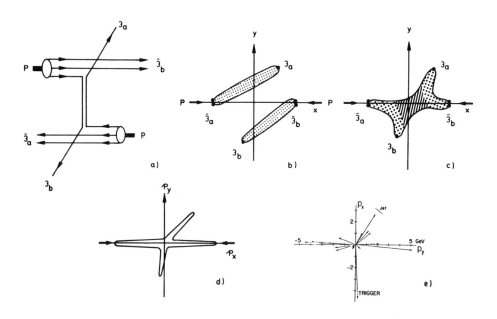

Fig.7.2. (a) Flow of color lines for elastic scattering of valence quarks; (b,c) possible configuration of color sources and flux tubes after the scattering for $r \gtrless 2$. Dotted and shaded regions denote triplet and octet strings, respectively; (d) final four jet event in momentum space; (e) a real large $p_\perp$ event. Momenta are projected into the scattering plane. From BFS /263/

This prediction was confirmed in 1973 by three experiments which studied proton-proton interactions at the CERN-ISR /8-10/. They measured a considerable excess of particles at transverse momenta above 1 GeV/c as compared to the extrapolation of low $p_\perp$ data. For meson production, the invariant cross section can be approximated by

$$E \frac{d^3\sigma}{dp^3} \sim p_\perp^{-8\ldots 9}$$

for

$$2 \lesssim p_\perp \lesssim 5\text{-}7 \text{ GeV}$$

After this discovery, a large amount of theoretical and experimental work concentrated on the following questions:

- Do high $p_\perp$ particles really result from a two body hard scattering process?
- Are the active partons identical with quarks and gluons, and is their interaction described by asymptotically free field theories, like QCD?
- Can one obtain further information on mechanisms of parton fragmentation and confinement, and is there a link between these events and normal inelastic hadron-hadron interactions where only low $p_\perp$ particles are produced?

The experimental investigation of events with high $p_\perp$ particles (in the following we consider the high $p_\perp$ regime as starting at $p_\perp$ around 2 GeV/c) is complicated by the complexity of the multijet final state. In each jet, the bulk of particles produced will have longitudinal momenta which are comparable to their momenta transverse to the jet axis. These particles are no longer aligned along the jet axis and, in general, cannot be attributed to the specific jet. Consequently, the momentum transfer q is no longer directly accessible to the experiment in contrast to lN charged current interactions.

The following discussion is concentrated on the last two of the questions mentioned above, which cannot be treated independently of each other.

This chapter is subdivided as follows. The large variety of experiments makes it necessary to discuss the main types of experiments. This will be done in the remainder of the introduction. In Sect.7.1 the predictions of the parton model are discussed on a very elementary level. In Sect.7.2 the main features of high $p_\perp$ events are compared with parton model predictions. The properties of the jets at high $p_\perp$ and of the spectator fragmentation are discussed in detail in Sects.7.3,4, respectively.

The data discussed in Sects.7.3,4 will mainly come from proton-proton interactions at the highest ISR energies, since there a reliable separation of the jets starts to be possible. In addition, the amount of final state interactions is minimized in proton-proton interactions as compared to hadron-nucleus collisions.

To avoid additional complications, the discussion of jet properties will be restricted to jets produced in hadron-hadron collisions. However, with the advent of more detailed and precise data, jet production off nuclei will be an important tool in studying the development of jets, because interaction with other nucleons tests the jet structure at early stages of confinement /250,251/.

Experiments investigating events with particles of large $p_\perp$ can be grouped into four main categories:

I) Experiments measuring inclusive single particle cross sections at highest $p_\perp$ and $\sqrt{s}$, with the aim of studying parton-parton interactions in the asymptotic region where corrections due to masses, finite transverse momenta, and higher twists are negligible. The tiny cross section requires large aperture spectrometers, which are usually realized by lead glass walls detecting high $p_\perp$ $\pi^{o'}$s.

II) Experiments comparing large $p_\perp$ cross sections for different particle species and different beam or target types. Particle ratios at large $p_\perp$ reflect the quantum numbers of the scattered partons and thus help to pin down the basic scattering mechanism. Such experiments commonly use magnetic spectrometers of very limited acceptance.

III) Experiments studying correlations between particles in events with a large $p_\perp$ secondary. Most of the results quoted in the following discussion come from this kind of experiment which can be set up in very different ways ranging from two small aperture spectrometers to 4π detectors. Since most of the recent detectors of type I) are equipped with a vertex spectrometer, they contribute to this field as well.

Experiments of type I) to III) use a large $p_\perp$, single-particle trigger. Their ability to measure parton-parton cross sections is restricted by the fact that one has to unfold an a priori unknown parton fragmentation function. A more reliable way to "detect" a scattered parton is to measure the whole jet, e.g., by determining the energy flux into a suited region of solid angle.

IV) Experiments measuring jet cross sections. These experiments are equipped with calorimeters measuring the amount of energy emitted at large angles in the cms. Although in principle being superior to other types of experiments measuring parton cross sections, these experiments suffer from the fact that even at ISR energies jets are far from being pencillike and that a cut in solid angle is a somewhat inefficient way to collect particles from one jet. Consequently, acceptance corrections are large, and there are ambiguities in interpreting the data.

Table 7.1 gives a list of large $p_\perp$ experiments, and summarizes their main characteristics.

7.1 Parton-Parton Scattering

In this section we discuss the kinematics and cross sections for parton-parton scattering.

As a first approximation and to simplify our presentation, we neglect parton masses and finite transverse momenta in the structure and fragmentation functions. Assume that the basic process is a two body interaction

$$q_1 q_2 \to qq' \tag{7.1}$$

where q_1, q_2 are partons inside the primary hadrons h_1, h_2 carrying momentum fractions x_1 and x_2, respectively.

In the cms of h_1 and h_2, the kinematics of the reaction is fully described by three "observables", the rapidities y and y' of the outgoing partons

$$y = \ln\left(\frac{E+p_\parallel}{m_\perp}\right) \cong -\ln \tan\left(\frac{\theta_{cms}}{2}\right) \tag{7.2}$$

and their transverse momenta $p_\perp = p'_\perp$ or transverse masses $m_\perp = m'_\perp = p_\perp$. x_1 and x_2 are given by

$$x_{1,2} = \frac{m_\perp}{\sqrt{s}} \left[\exp(\pm y) + \exp(\pm y')\right] \tag{7.3}$$

where s is the total cms energy.

For the Mandelstam invariants $\hat{s}, \hat{t}, \hat{u}$ of the subprocess we get

$$\begin{aligned}
\hat{s} &= (q_1 + q_2)^2 = x_1 x_2 s = 4m_\perp^2 \cosh^2\left(\frac{y-y'}{2}\right) \\
\hat{t} &= (q-q_1)^2 = -m_\perp^2 [1+\exp(y'-y)] \\
\hat{u} &= (q'-q_1)^2 = -m_\perp^2 [1+\exp(y-y')]
\end{aligned} \tag{7.4}$$

Table 7.1. Main characteristics of some large $p_\perp$ experiments

		Collaboration		Apparatus	Type of Exp.	Beam, Target	$\sqrt{s}$ [GeV]	Trigger Type	Trigger $p_\perp$ [GeV/c]	θ_{cms}	Some ref.
IRS		Saclay-Strasbourg	R 102	Single arm spectrometer	I)	pp	23-53	$h^\pm,\pi^0$	1-5	90°	/9/
		CERN-Columbia-Rockefeller (CCR)	R 103	Lead glass	I)	pp	53	π^0	2-8	90°	/10,252/
		British-Scandinavian (BS)	R 203	Single arm spectrometer	II)	pp	23-63	$\pi^\pm,k^\pm,p^\pm$	0-5	45°-90°	/8,253/
		Pisa-Stony Brook	R 801	Lead glass + szint. hodoscope	III)	pp	23-63	π^0	0-4	90°,17°,8°	/254/
		CERN-Columbia-Rockefeller-Saclay (CCRS)	R 105	Two arm spectrometer	I,III)	pp	23-63	$\pi^0,\pi^\pm$	2.5-8	90°	/255/
		CERN-Daresbury-Liverpool-Rutherford	R 205	Spectr. + szint. hodoscope	III)	pp	23-63	$\pi^\pm,k^\pm,p^\pm$	0-3	45°-90°	/256/
		Aachen-CERN-Heidelberg-Munich (ACHM)	R 701	Streamer chamber	III)	pp	45-63	π^0	0-4	53°-90°	/257/
		CERN-College de France-Heidelberg-Karlsruhe (CCHK)	R 407/ R 408	Split Field Magnet spectr.	III)	pp	53	$\pi^\pm,k^-,p$	1-4	20°,45°	/7,258,259/
		CERN	R 412	Split Field Magnet spectr.	III)	pp	53		1.5-4	90°	/260/
		British-French-Scandinavian (BFS)	R 413	Split Field Magnet spectr.	III)	pp	53	$\pi^\pm,k^\pm,p^\pm$	0.5-4	90°	/261-263/
		CERN-Columbia-Oxford-Rockefeller (CCOR)	R 108	Solenoid + lead glass	I,III)	pp	31-63	π^0	3-14	90°	/264/
		CERN-Saclay (CS)	R 702	Magnet spectr. + lead glass	I,III)	pp,pd,dd	53-63	π^0	3-15	90°	/265/
		Athens-Brookhaven-CERN Syracuse (ABCY)	R 806	Argon calorimeter	I,III)	pp	53-63	π^0	3-15	90°	/266/
		Annecy-CERN-College de France-Dortmund-Heidelberg-Warsaw (ACCDHW)	R 416	Split Field Magnet spectr.	III)	pp	23-63	π,k,p	> 5	45°	/266/
FNAL		Chicago-Princeton	E 100A	Single arm spectrometer	I,II)	pN	19-27	$\pi^\pm,k^\pm,p^\pm$	1-8	90°	/267/
		Chicago-Princeton	E 300	Single arm spectrometer	I,II)	pp,pN	19-27	$\pi^\pm,k^\pm,p^\pm$	1-7	90°	/268/
		Chicago-Princeton		Single arm spectrometer	II)	πp	19-24	$\pi^\pm,k^\pm,p^\pm$	1-6	60°-90°	/269/
		Brookhaven-Caltech-LBL	E 268/ E 350	γ-Calorimeter	II)	$\pi p,k^-p,p$ p	14-19	π^0	1-5	50°-110°	/270/
		Caltech-UCLA-Fermilab-Chicago-Indiana	E 260	Calorimeter + spectrometer	IV)	$\pi^\pm p,k^\pm p,pp,pN$	19	$\pi^\pm,k,p$ jet	1-4	60°-90°	/271/
		Fermilab-Lehigh-Pennsylvania-Wisconsin (FLPW)	E 395	Calorimeter	IV)	$\pi p,pp$	16-27	jet	2-4	50°-100°	/272,273/
		Columbia-FNAL-Stony Brook	E 494	Double arm spectrometer	III)	pN	19-27	$\pi^\pm,k^\pm,p^\pm$	1-5	90°	/274/
		FNAL-Michigan-Purdue	E 357	Double arm spectrometer	III)	pN	27	$\pi^\pm,k^\pm,p^\pm$	1-3,5	110°	/275,276/

where q_1, q_2, q, q' denote parton momenta. In analogy to the treatment of lepton-nucleon interactions we assume that the cross section $d^3\sigma/dydy'dp_\perp$ for the production of two partons q and q' factorizes as

$$d\sigma(h_1 h_2 \to qq') = \sum_{q_1, q_2} d\sigma(h_1 \to q_1) d\sigma(h_2 \to q_2) d\sigma(q_1 q_2 \to qq') \tag{7.5}$$

The first two terms are the well known hadron structure functions

$$d\sigma(h \to q) = G_h^q(x) dx/x \tag{7.6}$$

For the cross section we write

$$d\sigma(q_1 q_2 \to qq') = \frac{d\sigma^{12}}{d\hat{t}}(\hat{s}, \hat{t}) d\hat{t} \tag{7.7}$$

Using $(dx_1/x_1)(dx_2/x_2)d\hat{t} = dydy'dp_\perp^2$ (7.7) finally yields (for simplicity, we omit the sum over q_1, q_2),

$$\frac{d\sigma}{dydy'dp_\perp^2} = G_{h_1}^{q_1}(x_1) G_{h_2}^{q_2}(x_2) \frac{d\sigma}{d\hat{t}} \tag{7.8}$$

Writing σ as

$$\frac{d\sigma}{d\hat{t}}(\hat{s}, \hat{t}) = \alpha_s \hat{s}^{-n} f\left(\frac{\hat{t}}{\hat{s}}\right) = \alpha_s \hat{s}^{-n} f(\hat{\theta}) \tag{7.9}$$

we obtain the scaling law

$$\frac{d\sigma}{dydy'dp_\perp^2} = \alpha_s F(y, y', x_\perp) p_\perp^{-2n} \quad \text{(with } x_\perp = 2p_\perp/\sqrt{s}) \tag{7.10}$$

where the angular dependence of $d\sigma/d\hat{t}$ and the structure functions are summarized in F, and the term $p_\perp^{-2n}$ reflects the $\hat{s}$ dependence of $d\sigma/d\hat{t}$ at fixed angle $\hat{\theta}$ [note that scale breaking effects are neglected in (7.10)!]

Using dimensional arguments, n is given by

$$\left(\frac{d\sigma}{dt}\right) = (\alpha_s)\left(\frac{1}{\hat{s}}\right)^n \tag{7.11}$$

which yields n = 2 for theories with a dimensionless coupling constant.

The cross sections (7.9) and (7.10) refer to the production of large $p_\perp$ partons or jets. The single-particle cross section σ^h is given by a convolution of σ over the parton fragmentation function $D(z)$

$$E\frac{d^3\sigma^h}{dp^3}(p_\perp) = \int\frac{dz}{z^2} E\frac{d^3\sigma}{dp^3}\left(\frac{p_\perp}{z}\right) D(z) \tag{7.12}$$

In order to enable simple analytical calculations, we parameterize σ locally as $E(d^3\sigma/dp^3) \cong A p_\perp^{-k}$ and choose a rather general ansatz for $D(z)$

$$D_q^h(z) = (B/z)(1-z)^m + L + T\delta(1-z) \tag{7.13}$$

The δ term takes into account the possible emergence of stable hadrons from certain subprocesses. From (7.12) and (7.13) we obtain for the ratio of inclusive single particle and jet cross sections /277-279/

$$(E\frac{d^3\sigma^h}{dp^3})/(E\frac{d^3\sigma}{dp^3}) = 2B\frac{m!(k-3)!}{(k+m-2)!} + \frac{L}{(k-1)} + T \tag{7.14}$$

Within our approximation the single-particle spectrum has the same slope in $p_\perp$ as the distribution of jets, however, the absolute yield is much smaller. For example, by choosing the standard quark fragmentation function with $m \cong 2-3$, $L \ll 1$ and $T = 0$, one gets ratios of the order 10^{-2}.

This fact is easy to explain. For a large k and a given $p_\perp^h$, it is less "expensive" to pick up a hadron at $z \cong 1$ from a jet with $p_\perp \cong p_\perp^h$ than to use a slow fragment of a parton at very large $p_\perp$. As k increases, the mean z of the trigger hadron approaches 1

$$\left\langle\frac{1}{z}\right\rangle = \frac{\langle p_\perp\rangle}{p_\perp^h} = \frac{\int\frac{dz}{z^2} E\frac{d^3\sigma}{dp^3}\left(\frac{p_\perp^h}{z}\right) D(z)\frac{p_\perp^h}{z}}{\int\frac{dz}{z^2} E\frac{d^3\sigma}{dp^3}\left(\frac{p_\perp^h}{z}\right) D(z) p_\perp^h} = \frac{\left[Bm!(k-4)! / (k+m-3)! + \frac{L}{k-2} + T\right]}{\left[Bm!(k-3)! / (k+m-2)! + \frac{L}{k-1} + T\right]} \tag{7.15}$$

and σ^h drops. Experiments triggered on a single hadron at large $p_\perp$ thus select a special type of parent jets, those consisting essentially of one very fast fragment. This effect is known as the trigger bias /278,279/. Experiments triggered by jets of large $p_\perp$ are not subjected to this bias. However, a similar bias is introduced if the solid angle covered by the jet detector is comparable to the jet size. Then the trigger condition enhances narrow, well collimated jets /273/.

Another important consequence of the trigger bias is that for a mixture of different parent partons the species with the flattest fragmentation function is favored by the single particle trigger condition.

Of course, only the jet containing the trigger particle is affected by the trigger bias. The recoiling parton in the opposite azimuthal hemisphere decays unbiased within the limits imposed by energy-momentum conservation. The fragmentation of this second "away" jet is usually described in terms of a variable x_E referring to the trigger momentum $p_\perp^h$

$$x_E = p_\perp^{h'}/p_\perp^h \tag{7.16}$$

where h' is a hadron in the away jet. Neglecting transverse momenta in the parton fragmentation, x_E is related to the "correct" scaling variable $z^{h'}$ referring to the momentum of the away jet

$$z^{h'} = x_E z^h \tag{7.17}$$

(Note that x_E may exceed 1.) The particle density $(1/\sigma)(d\sigma/dx_E)$ is obtained from

$$\frac{1}{\sigma}\frac{d\sigma}{dx_E} = \frac{\int \frac{dz}{z} E\frac{d^3\sigma}{dp^3}(\frac{p_\perp^h}{z})D(z)D'(x_E z)}{\int \frac{dz}{z^2} E\frac{d^3\sigma}{dp^3}(\frac{p_\perp^h}{z})D(z)} \tag{7.18}$$

$D(z)$ and $D'(z)$ are fragmentation functions of the towards and the away jets, respectively. Two points should be kept in mind. Since $d\sigma/dx_E$ depends both on $D(z)$ and on the inclusive jet cross section $d^3\sigma/dp^3$, scaling of the fragmentation functions does not necessarily imply scaling of $d\sigma/dx_E$ in x_E. Furthermore, the dependence of $d\sigma/dx_E$ on the trigger-side fragmentation function $D(z)$ may induce correlations between the towards and the away jet. Compare, for example, events where scattered u- or d-quarks create $\pi^\pm$ and K^- trigger particles. Since $D_{u,d}^{K^-}$ falls steeper in z than $D_{u,d}^\pi$, the mean momentum of the towards jet is larger for the K^- trigger, and $d\sigma/dx_E$ is flatter. Care is needed not to confuse such kinematical effects with dynamical correlations due to the scattering mechanism itself.

To complete this discussion let me quote the relations concerning the two spectator jets at low $p_\perp$. The energy available to the spectators is reduced when compared to the total cms energy. The sum of the energies or momenta of the spectator jets is

$$E' = \sqrt{s} - m_\perp(\cosh y + \cosh y')$$
$$p' = -m_\perp(\sinh y + \sinh y') \qquad (7.19)$$

Hence, the usual Feynman variable $2p_\parallel/\sqrt{s}$ is no longer suitable to describe the fragmentation of spectators. The most natural choice is to use a reduced energy $\sqrt{s}'$ for the system of the two spectator jets

$$\sqrt{s}' = (E'^2 - p'^2)^{1/2}$$

The appropriate scaling variable is then the reduced longitudinal momentum of a secondary in the spectator system /259/

$$x' = 2p'_\parallel/\sqrt{s}' \qquad (7.20)$$

7.1.1 The QCD Approach

In a hard scattering picture, the favored candidates for the active partons are quarks and gluons, their interactions governed by QCD. The basic ingredients of the model are shown in Figs.7.3-5. The main subprocesses are quark-quark, quark-gluon, and gluon-gluon scattering with the cross sections /280,281/ given by

$$\frac{d\sigma}{d\hat{t}}(q_1 q_2 \to qq') = \pi \alpha_s^2(Q^2) \frac{f(\hat{\theta})}{\hat{s}^2} \delta_{q_1 q} \delta_{q_2 q'} \qquad (7.21)$$

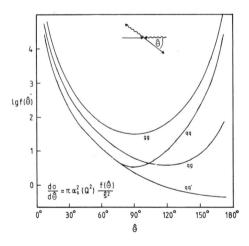

Fig.7.3. Angular dependence of parton-parton cross sections in first order of QCD /280,281/

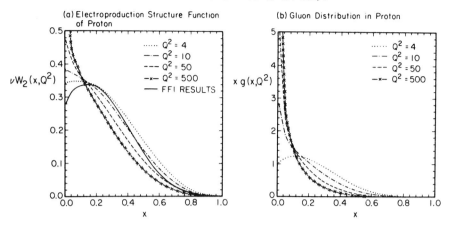

Fig.7.4. Typical parameterization of quark and gluon structure functions of the proton for various Q^2 /282/

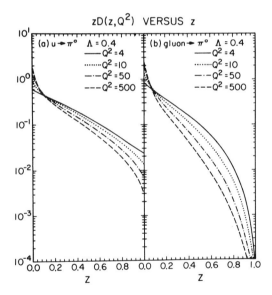

Fig.7.5. As Fig.7.4, but for fragmentation functions

whose angular dependence $f(\hat{\theta})$ is displayed in Fig.7.3. The δ function in (7.21) refers to parton flavors only, color factors are included in $f(\hat{\theta})$. Since the effective color charge of a gluon is 3/2 times that of a quark, we get (for forward scattering)

$$\sigma_{\text{gluon gluon}} > \sigma_{\text{gluon quark}} > \sigma_{\text{quark quark}} \qquad (7.22)$$

As factorization holds in QCD at the leading log level /146-148/, the parton structure functions and fragmentation functions are identical to those measured in lepton-nucleon reactions up to corrections $O(\alpha_s)$ which mainly affect the distribution of sea quarks. The Q^2 evolution of $G(x,Q^2)$ and $D(z,Q^2)$ is governed by the Altarelli-Parisi equations (5.23,30). However, the large mass scale Q^2 characterizing the hard scattering is not uniquely determined in a leading log calculation. The naive choice $Q^2 = -\hat{t}$ is not fully adequate to describe a quark-gluon compton effect, e.g. Forms used in the literature are /281-284/

$$Q^2 = \frac{2\hat{s}\hat{t}\hat{u}}{(\hat{s}^2+\hat{t}^2+\hat{u}^2)}, \quad Q^2 = (\hat{s}\hat{t}\hat{u})^{1/3}, \quad Q^2 = p_\perp^2$$
$$Q^2 = -\hat{t}, \quad Q^2 = \hat{s}, \quad Q^2 = (\hat{s}-\hat{t}-\hat{u})/3 \qquad (7.23)$$

resulting in uncertainties of the single-particle cross section of the order of 20-30%.

Typical examples for structure functions and parton fragmentation functions are shown in Figs.7.4-5. Although the basic cross section (7.21) scales as $\hat{s}^{-2}$, scale breaking effects increase the effective power in $p_\perp$ of the single particle cross section by about 2 units yielding /285/

$$E(d^3\sigma/dp^3)_{x_\perp} \sim p_\perp^{6...7} \qquad (7.24)$$

in contrast to the naive expectation $p_\perp^4$ (Fig.7.6)

Whereas the distribution of quarks in a nucleon is determined rather precisely from lN scattering experiments, the gluon density is accessible experimentally only through scaling violations in the distributions of sea quarks and is not very well determined. Nevertheless, most authors agree that at moderate $Q^2 \simeq 2$-4 GeV^2 the gluon distribution in the proton more or less coincides with the counting rule prediction

$$G_p^g(x) \sim (1-x)^{4-5} \qquad (7.25)$$

with the normalization $\int dx G_p^g(x) = 0.5$. Based on theoretical prejudices, the gluon fragmentation function is typically chosen a factor $(1-z)$ steeper than the quark fragmentation function.

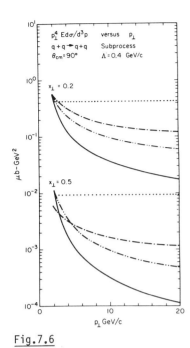

Fig. 7.6

Fig. 7.6. QCD predictions for the cross section $p_\perp Ed^3\sigma/dp^3$ for $pp \to \pi^0 X$ at $90°$ in the cms, at ISR energies /285/. (····) "QCD" without scale breaking. (·-·-·) with running coupling constant $\alpha_s(Q^2)$. (··-··-) with running coupling and non-scaling structure functions. (———) QCD including all scale breaking effects

Fig. 7.7. QCD prediction for the contribution of different subprocesses to the cross section $pp \to \pi X$ at $\sqrt{s} = 52$ GeV compared to experimental data /286/

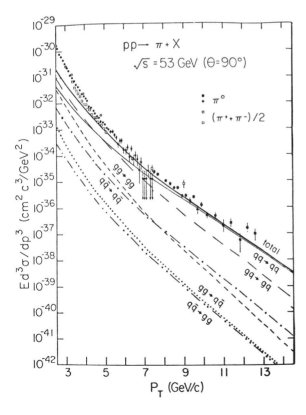

Fig. 7.7

Figure 7.7 shows a QCD prediction for the inclusive pion cross section at large angles for the reaction pp → π+X, subdivided into different subprocesses. The calculation is in fair agreement with the data for $p_\perp \gtrsim 5$ GeV/c. At lower $p_\perp$ the theory falls considerably below the data, indicating that additional scale breaking or higher twist effects have to be taken into account. Such corrections could arise from a nonnegligible transverse momentum of partons in the nucleon or from additional subprocesses involving nonelementary constituents with form factors, as are postulated by the constituent interchange model (CIM) /209,213,215,287-289/.

7.1.2 Parton Transverse Momentum

In the naive parton model, the partons q_1 and q_2 move collinear with their parent hadrons h_1 and h_2. There is, however, considerable evidence, particularly from dilepton production experiments, that the partons can have a sizeable transverse momentum. It has been pointed out that these effects are very important in large $p_\perp$ calculations where the steeply falling cross section is sensitive to the transverse configuration of the initial state partons /7,290,291/.

Although the qualitative effect of parton transverse momentum is fairly well understood (triggering on a large $p_\perp$ hadron favors configurations where both the active constituents move in the transverse direction of the trigger hadron, thereby reducing the effective $\hat{t}$ and enhancing the cross section), there is no common consensus as how to incorporate these effects into the calculation of cross section, and quantitative predictions show major discrepancies /282,283,292-295/.

Two components contribute to the parton transverse momentum $k_\perp$: a primordial $k_\perp$ inherent to the initial wave function, and a component created through hard bremsstrahlung during the interaction (Fig.7.8). Naively, one expects the primordial $k_\perp$ to be of the order of 300 MeV, whereas the hard component grows as $Q^2/\ln(Q^2/\Lambda^2)$. To circumvent problems arising in the calculation of higher-order QCD diagrams /296-298/, most authors parameterize both components by one effective $k_\perp$ distribution. Choosing a mean $k_\perp$ of 850 MeV, independent of x and Q^2, the theory can be tuned to agree with data over the whole range of $p_\perp$ (Fig.7.9).

The phenomenological way of including parton $k_\perp$ raises several new problems. Figure 7.8b can be regarded either as a quark-quark scattering, with the effect of the bremsstrahlung gluon being parameterized by the parton $k_\perp$, or as a quark-gluon compton effect. The same process appears to be counted twice.

In the context of parton $k_\perp$, another point becomes evident: it is no longer possible to keep all partons on shell /189,190/. Two ways are followed in the literature: either the active partons are massless and kept on shell /282/, with the fate of the spectators being neglected, or it is argued that the spectator essen-

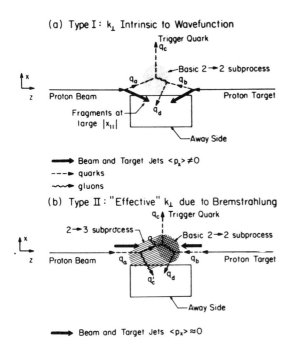

Fig.7.8a,b. Effect of a parton transverse momentum in parton elastic scattering processes (a) primordial transverse momentum intrinsic to the wave function (b) transverse momentum due to gluon bremsstrahlung. The trigger condition (a large $p_\perp$ hadron) selects configurations where the parton transverse momentum is aligned with the final hadrons momentum /282/. Note that the second diagram could be counted either as quark-quark scattering with a bremsstrahlung gluon, or as a quark-gluon scattering, where the gluon recoils against another quark

tially carries the initial hadron mass, since it is subjected only to confining forces which have an appreciable effect only after a rather long time /190,295/. Consequently, the active partons are off shell (6.2). This procedure has the advantage that the pole of scattering amplitude at $\hat{t} = 0$ moves into the unphysical region and cannot be reached. In the case of on shell partons, an arbitrary regularizing mass $\hat{t} \to \hat{t}-M^2$ has to be introduced to avoid divergencies /7,282/.

In the spirit of a preconfinement model, none of these methods are fully adequate. Here the active partons become more and more off shell through successive emission of gluons, reaching a mean momentum squared of the quark line at the hard vertex up to $O(Q^2)$. After the scattering, the active parton initiates a "parton shower" and successively cascades down to the mass shell. Recent works which have begun to deal with this problem /301/, choose the parton mass as the mean mass

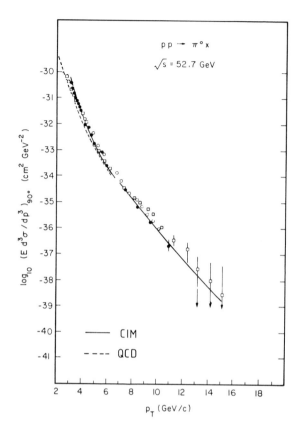

Fig.7.9. Prediction of QCD /285/ and CIM /288/ for the single-particle cross section at large $p_\perp$ at $\theta_{cms} \cong 90°$/264-266/. The QCD model uses a mean parton transverse momentum of 850 MeV/c /285/. Above $p_\perp \cong 7$ GeV/c, both data and CIM model refer to the sum of inclusive π^0 and direct γ spectra

of a Sterman-Weinberg jet of corresponding momentum, with the result that the QCD jet cross section is fairly insensitive to such manipulations.

Anyhow, all these ambiguities in the interpretation of the QCD calculus concern mainly the region below $p_\perp \cong 5$ GeV, and it is commonly agreed upon that the inclusion of parton transverse momentum improves the agreement between theory and experiment in the medium $p_\perp$ range /7,282,292-295/.

7.1.3 Constituent Interchange Model (CIM)

The QCD interpretation of large $p_\perp$ phenomena requires a rather large parton transverse momentum. The phenomenological representation of these fluctuations by an effective $k_\perp$, instead of a sum over Feynman diagrams, raises various problems, part of which can be traced back to the fact that at moderate $p_\perp$, the lifetime of a large $k_\perp$ fluctuation is of the same order as the timescales set by the hard interaction. In this case, effects due to the coherence of the initial wave function have to be taken into account, resulting in subprocesses involving more than two quark partons.

The CIM model /209,212,213,287-289/ assumes that particle production in the region below $p_\perp$ = 5 to 7 GeV/c is governed by such "higher twist" or "natural" mechanisms. Typical subprocesses and their $p_\perp$ dependence are

$$\left.\begin{array}{l} qM \to qM \\ q\bar{q} \to MM \end{array}\right\} \quad p_\perp^{-8}$$
$$qB \to qB \quad p_\perp^{-12} \tag{7.26}$$

The labels M, B and q denote partons with meson, baryon, or quark quantum numbers, respectively, and are treated as having a negligible primordial $k_\perp$.

It has been argued /288,302,304/ that a systematic treatment of all hard scattering processes including both elementary quark and gluon contributions and higher-order constituent interchange processes is given by the "hard scattering expansion" /303/, which accounts for gluon corrections to the basic QCD process, constituent transverse momenta, higher-twist effects, etc. For example, inclusive meson cross sections are represented by

$$E\frac{d^3\sigma}{dp^3}(h_1 h_2 \to hX) = \underbrace{\frac{F(x_\perp,\theta)}{p_\perp^4}}_{\substack{qq \to qq \\ qg \to qg}} + \underbrace{\frac{G(x_\perp,\theta)}{p_\perp^6}}_{\substack{gM \to gM \\ qg \to qM}} + \underbrace{\frac{H(x_\perp,\theta)}{p_\perp^8}}_{\substack{qM \to qM \\ qq \to MM}} + \cdots \tag{7.27}$$

The intrinsic transverse momentum of the partons in (7.27) is small and reflects nonleading contributions. The validity of this expansion has been proven rigorously in a ϕ^3 toy theory /303/; it is presently not clear if it is appropriate to asymptotically free theories such as QCD.

The $p_\perp^{-4}$ terms in (7.27) correspond to the asymptotic QCD predictions; the $p_\perp^{-8}$ terms represent the classical CIM mechanisms. The $p_\perp^{-6}$ processes are suppressed by a subtle cancellation related to gauge invariance /302/.

The normalization of the different processes can be derived from meson form factors and elastic scattering cross sections /305/. The resulting cross section is in fair agreement with data (Fig.7.9), the dominant mechanism below $p_\perp \simeq$ 5-7 GeV being $qM \to qM$.

From the experimentalists point of view, QCD and pure CIM differ mainly in three points:

I) In the CIM model, the "jet" containing the trigger particle is formed by decay products of a meson resonance.

II) Mesons which do not contain one of the incident valence quarks are likely to be produced via $q\bar{q} \to MM$. For $pp \to K^-X$ one obtains /302/

$$\frac{\sigma(qq \to K^-M)}{\sigma(qM \to K^-q)} \approx \frac{0.15}{(1-x_\perp)^2} \quad \text{for } \theta \simeq 90° \text{ and } x_\perp > 0.15 \tag{7.28}$$

Compared to large $p_\perp \pi^-$ triggers, which in the CIM are almost entirely due to $qM \to \pi^- q$, the away jet for a K^- trigger should be different, and its contents of K^+ should be enhanced in contrast to QCD where no strong flavor correlation exists between the two jets at large $p_\perp$.

III) If one of the colliding hadrons contains a valence antiquark, the $q\bar{q} \to MM$ subprocess will considerably enhance the cross section at large $x_\perp$, as compared to QCD predictions.

7.2 General Characteristics of High $p_\perp$ Events

Scattering of quark or gluon partons was shown to account for single-particle yields at large $p_\perp$, once higher-order and higher twist effects are taken into account. In this section we shall discuss further evidence for a basic two-body scattering process, using two sets of data which are insensitive to details of parton fragmentation. The mechanisms of fragmentation and the properties of jets in large $p_\perp$ events will be discussed in Sects.7.3,4.

7.2.1 Particle and Beam Ratios

Ratios of single particle or jet cross sections at large $x_\perp$ for different particle species, beam, and target types essentially test the ratio of structure functions of the interacting partons. They have the great advantage that both theoretical and experimental uncertainties tend to cancel.

The ratio of jet to single-particle cross sections (Fig.7.10) tests the idea that scattered partons carry color and hence must fragment. This would yield a tremendous ratio for $\sigma^{Jet}(p_\perp)/\sigma^h(p_\perp)$ of the order of 10^3 at large $x_\perp$ /282/ (7.14). The measured ratio is in good agreement with QCD predictions. For pure CIM processes which directly produce color-singlet mesons, the jet cross section is lowered by about one order

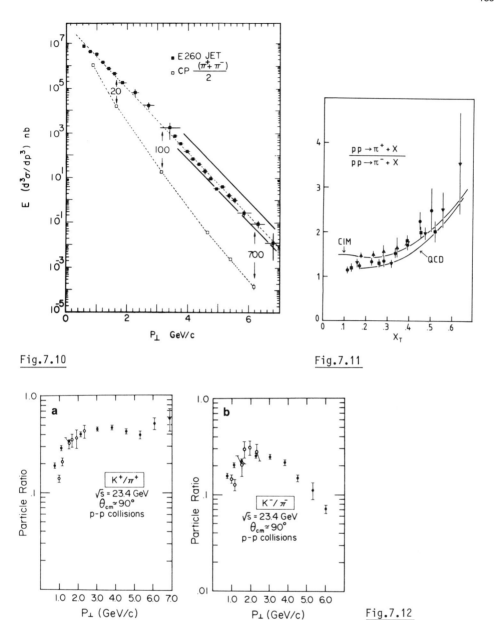

Fig.7.10

Fig.7.11

Fig.7.12

Fig.7.10. Jet cross section at $\sqrt{s}$ = 19.4 GeV compared with the single-particle cross section /271/. The full lines refer to a QCD calculation for fixed jet energy and fixed jet momentum, respectively

Fig.7.11. Particle ratio $pp \to \pi^+/\pi^- + X$ at $\theta_{cms} \cong 90°$ from /268/. (▲) $\sqrt{s}$ = 19 GeV; (■) $\sqrt{s}$ = 23 GeV; (●) $\sqrt{s}$ = 27 GeV and /353/; (▼) $\sqrt{s}$ = 19 GeV compared to QCD /282/ and CIM /287/ predictions

Fig.7.12a,b. Particle ratios $pp \to k^+/\pi^+ + X$ and $pp \to k^-/\pi^- + X$ at $\theta_{cms} \cong 90°$ vs $p_\perp$ for $\sqrt{s}$ = 23.4 GeV (○): /253/, (●) /268/

of magnitude /302,304/ as compared to QCD. In the subprocess expansion (7.32) combining QCD and CIM graphs, the relative abundance of processes can be adjusted such that the major contributions to the jet cross section come from QCD graphs, whereas CIM terms dominate single-particle production, favored by the trigger bias. Within present experimental errors /271,273,307/ such a combination is indistinguishable from pure QCD.

Figures 7.11,12 show π^+/π^-, k^+/π^+ and k^-/π^- ratios at large $p_\perp$ for proton-proton collisions. In almost any hard scattering model referring to quark partons, π^+, k^+ and π^- mesons at large $x_\perp$ or x contain an u and d valence quark, respectively from one of the incident hadrons. Consequently, the π^+/π^- ratio at large $x_\perp$ reflects the ratio $1/(1-x)$ of u- and d-quark structure functions, while k^+/π^+ should be constant for $x \to 1$, and k^-/π^- will drop with increasing $p_\perp$. Data show all these features, proving that the standard valence quarks are involved in large $x_\perp$ particle production.

The same argument shows that the beam ratio $pp \to$ Jet + X/$\pi p \to$ Jet + X falls with increasing $x_\perp$, since the x distribution of valence quarks is flatter in a pion, as compared to a proton, simply because momentum is shared by only two valence quarks. Again data agrees with the QCD ideas (Fig.7.13). The CIM predictions, where the process $qM \to qM$ is considerably enhanced due to the incident meson, falls below the data.

Beam ratios also offer a simple way to decide whether the underlying process is a scattering ($\hat{t}$ channel) or fusion ($\hat{s}$ channel) mechansim. The ratio $pp \to$ Jet + X/ $p\bar{p} \to$ Jet + X should be about 1 in the first case and small compared to unity else. The conclusion from Fig.7.14 is evident.

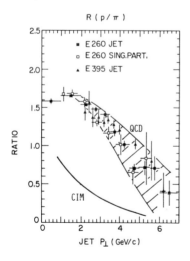

Fig.7.13. Beam ratio $pp \to \pi^0 + X/\pi p \to \pi^0 + X$ and $pp \to$ Jet + X/$\pi p \to$ Jet + X /271,272/ compared to QCD /306/ and CIM /287/ predictions. The shaded area gives the QCD prediction for jets (upper boundary) and single particles (lower boundary). The CIM prediction refers to single particles

Finally, one can try to study the angular dependence of the basic scattering process by considering the particle ratio π^-/π^+ in $\pi^- p$ collisions as a function of the cms angle. If $d\sigma/d\hat{\theta}$ is peaked forward, one expects that π^-/π^+, measured at forward angles in the pion hemisphere, is large and increases with $p_\perp$ since the incoming pion contains two negative, but no positive valence quarks.

Figure 7.15 compares very recent experimental results with QCD predictions for cms angles of $90°$, $77°$, and $60°$. The expected tendency is not observed, except perhaps for the two points at highest $p_\perp$ and $\theta_{cms} \cong 60°$. A possible explanation for the discrepancy may be that the gluon contribution is underestimated since the gluon fragmentation function is chosen too steep. In that case the expected behavior shows up only at larger $p_\perp$ or smaller θ_{cms}. However, it is obvious that these data, if confirmed by future experiments, pose a serious problem for QCD.

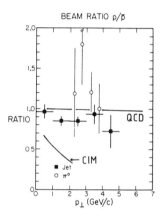
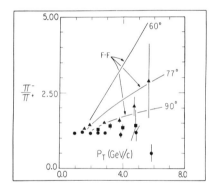

Fig.7.14. Beam ratio $pp \to Jet + X/p\bar{p} \to Jet + X$ and $pp \to \pi^0 + X/p\bar{p} \to \pi^0 + X$ vs $p_\perp$ at $\sqrt{s} = 19$ GeV /270,271/ compared to QCD /306/ and CIM /287/ predictions

Fig.7.15. Particle ratio $\pi^- p \to \pi^- + X/\pi^- p \to \pi^+ + X$ as a function of $p_\perp$ for cms angles of $60°$ (▲), $77°$ (■), and $90°$ (●), at $\sqrt{s} = 19.4$ GeV, compared to QCD predictions /269/

7.2.2 Structure of Large $p_\perp$ Events

The inspection of particle and beam ratios supports the idea that in events with a secondary at large $x_\perp$, a valence quark is struck out of one of the incident hadrons. What remains to be proven is that this happens by a two-body process, i.e., that a large $p_\perp$ event contains four jets of particles, two at large $p_\perp$ and two low $p_\perp$ jets made of spectator fragments.

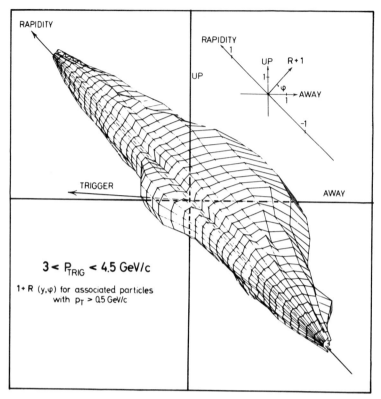

Fig.7.16. Ratio of particle densities in large $p_\perp$ events and in normal inelastic events as a function of y and ϕ. From BFS /262/

Figure 7.16 shows the ratio of particle densities observed in pp collisions with a large $p_\perp$ particle at $y^t \simeq 0$ to those observed for normal inelastic events /262/ as a function of the rapidity y and the azimuthal angle ϕ. The observed structure is not too far from what one expects in a hard scattering model. Near the trigger hadron, the particle density is enhanced, indicating the existence of a "towards" jet. The increase in density opposite in azimuth to the trigger is commonly interpreted as due to the away jet. Since the momenta of the colliding partons vary, particles from the away jet are smeared out over a wide range in y, when averaged over many events.

Contributions from the two jets at large $p_\perp$ are more clearly seen in Fig.7.17. In the central rapidity region, the particle density peaks at $\phi = \phi_{trigger} = 0$ and at $\phi = \pi$. With increasing $p_\perp$ of the secondaries, the flat background of spectator fragments diminishes, and the jets are better collimated in angle /264/.

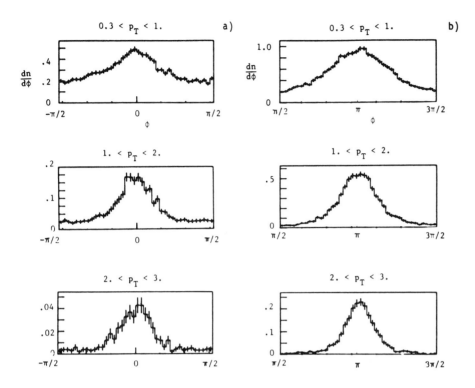

Fig.7.17a,b. Density of additional secondaries produced in events with a large $p_\perp$ (>7 GeV/c) π^0 at $\phi = 0$ as a function of azimuthal angle for different $p_\perp$ of the secondaries. From CCOR /264/

Figures 7.18,19 demonstrate that the increase of particle density in the towards region $\phi \cong \phi^t$ follows the trigger rapidity, whereas the density of away secondaries stays symmetrical around $y = 0$ /7/. Since particles of both like and opposite charge as the trigger show a narrow correlation, the effect cannot be entirely due to the decay of low-lying resonances like ρ /258,316/.

Consider now the azimuthal region opposite to the trigger. A comparison of the $p_\perp$ spectrum of particles with the inclusive spectrum proves that not only the number of particles, but their mean $p_\perp$ as well are increased (Fig.7.20 /255/). If the excess observed in Fig.7.20 is attributed to the "away" jet, two particles in this region are expected to show strong correlations in rapidity since both are more or less aligned along the jet axis. Such a correlation, whose strength increases with the $p_\perp$ of particles, is in fact observed (Fig.7.21 /265/). The shape of the correlation function is similar to that observed for towards particles. No correlation is seen between the trigger rapidity and away particles. Regarding the strength of the two-particle correlation function $C(y_1, y_2)$ at $y_1 \cong y_2$ as a measure for the pro-

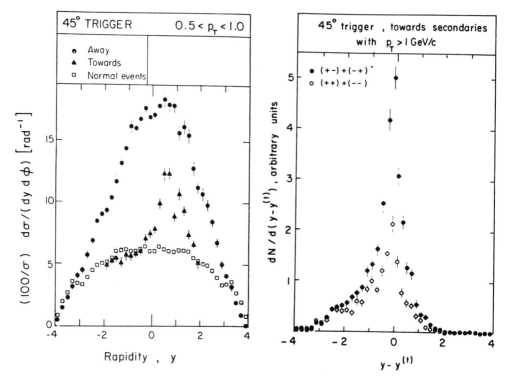

Fig.7.18. Rapidity distributions of secondaries integrated over $p_\perp$ from 0.5 to 1.0 GeV/c. Shown are the distributions away ($\phi \simeq \phi^t + 180° \pm 40°$) and towards ($\phi \simeq \phi^t \pm 25°$) secondaries for a trigger particle at $y \simeq 0.8$-1.0 and of particles produced in normal events. The mean $p_\perp$ of the trigger hadron is 2.4 GeV/c. From CCHK /258/

Fig.7.19. Same data sample as in Fig.7.18. Shown is the distribution of toward secondaries with $p_\perp > 1$ GeV/c in rapidity relative to the trigger rapidity. Full circles represent secondaries of charge opposite to the trigger charge, open circles refer to secondaries of like charge as the trigger. From CCHK /258/

bability to find the away jet at y, one concludes that its rapidity range is limited to $|y| < 2$-3 at ISR energies (Fig.7.22 /258/). Note further that the correlation shown in Fig.7.22 is practically independent of the trigger rapidity.

In hard scattering models, one expects the two jets to be coplanar. It is, however, not clear, if the back-to-back structure of jets seen in Fig.7.17 is not merely a consequence of momentum conservation. This question has been investigated by studying events with two π^0's of large $p_\perp$ /266/. Figure 7.23 shows their difference in azimuth for events having a fixed value of E_T. E_T is defined as the sum of the pion transverse momenta, plus the $p_\perp$ of another object necessary to ensure the $p_\perp$ balance, $\underline{p}_{\perp X} = -\underline{p}_{\perp \pi^0} - \underline{p}_{\perp \pi^0}'$. For uncorrelated emission of particles as described by

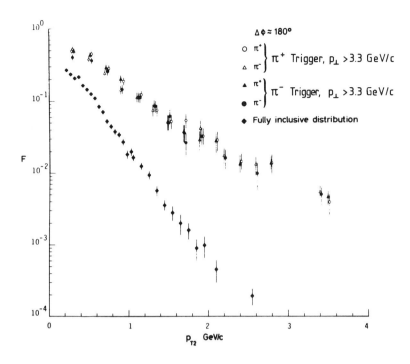

Fig.7.20. Density of particles produced opposite in ϕ to the trigger particle as a function of the transverse momentum of secondaries compared to the inclusive cross section. From CCRS /255/

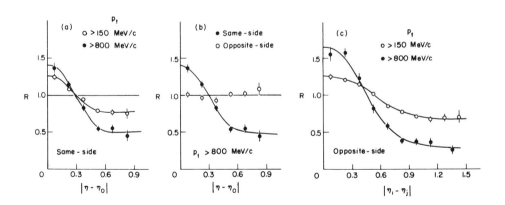

Fig.7.21a-c. Pair correlation functions for rapidities of towards and away secondaries. (a) Between the trigger π^0 and charged particles on the same side, (b) between the trigger π^0 and charged particles with $p_\perp > 800$ MeV/c, (c) between charged particles in the away region. From CS /265/

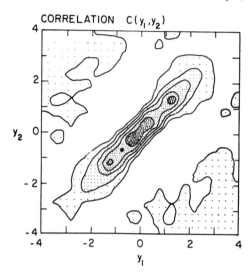

Fig.7.22. Contour plot of correlation between away particles with $p_\perp > 800$ MeV/c for a mean trigger $p_\perp$ of ~2.5 GeV/c. From CCHK /258/

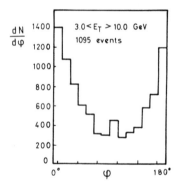

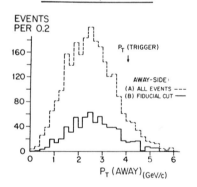

Fig.7.23. Distribution of azimuthal angle between two large $p_\perp$ π^0's, for $8 \leq E_\perp \leq 10$ GeV/c. From ABCY /266/

Fig.7.24. Transverse momentum detected in the away side calorimeter when triggering on a jet with $3.95 < p_\perp < 4.2$ GeV/c. From E 395 /272/

the UJM (Chap.3), the production probability depends only on $E_\perp$ and not on the angle between π^0s. Figure 7.23 proves that even at fixed $E_\perp$ the distribution peaks at $\Delta\phi \simeq 0°$ and at $\Delta\phi \simeq 180°$.

Finally, is there exactly one away jet in each event, and does it compensate the whole $p_\perp$ of the trigger particle? Figure 7.24 shows the momentum distribution of the away jet as obtained from a calorimeter experiment /272,273/. The discrepancy between the trigger $p_\perp$ and the mean momentum of the away jet cannot be explained by the limited acceptance of the calorimeter /307/. A natural explanation is given by assuming a parton transverse momentum of ~1 GeV/c. In this case, the transverse motion of the active partons is aligned along the trigger direction, with the recoil being taken by the spectators. This interpretation is supported by the investgation of spectator fragmentation /7,261/ (Sect.7.4).

In Fig.7.25 the frequency of reconstructed away jets with $p_\perp > 1.5$ GeV/c is plotted as a function of the trigger $p_\perp$. As a reference, two curves from a Monte Carlo simulation of the jet reconstruction procedure are shown, one for zero parton transverse momentum, and one assuming a momentum difference between trigger and away jet of ~0.8 GeV/c due to parton $k_\perp$. The away jet was chosen to resemble those observed in e^+e^- annihilations. The result for nonzero parton $k_\perp$ supports the idea that an away jet is present in each large $p_\perp$ event.

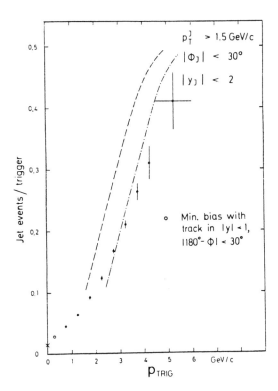

Fig.7.25. Frequency of reconstructing a jet on the away side. The curves refer to a Monte Carlo simulation assuming either perfect momentum balance the jets (----) or an imbalance of 0.8 GeV/c due to parton transverse momenta (-·-·-). From BFS /263/

To conclude: correlation data are fully consistent with the assumption of an underlying two-body hard scattering process. Note, however, that this does not necessarily give new information compared to the investigation of inclusive cross sections. It has been pointed out /308,309/ that in an uncorrelated jet model (Sect. 3.1) whose matrix element is chosen power behaved in $p_\perp$, a large transverse momentum is likely to be compensated by a single particle or cluster. If such a model is used to describe parton thermodynamics, all kinematical features of large $p_\perp$ events are reproduced once the single parton matrix element is matched to describe the inclusive spectrum.

7.3 The Jets at Large $p_\perp$

The investigation of properties of the two jets at large $p_\perp$ is concentrated on the following questions.

- Does factorization hold, i.e., does a quark fragment similarly in e^+e^- annihilations or in such a complex environment of color sources as is given in large $p_\perp$ reactions?
- Are the fragmentation functions of the towards and away jets identical? According to QCD, the away side parton is likely to be a gluon, whereas the towards parton is more often a quark /282/.
- Is there any evidence for scale breaking effects or for gluon bremsstrahlung?
- Is there a nontrivial correlation between the flavors of the towards and the away parton? Such a correlation would be a hint that additional mechanisms are active besides the basic QCD graphs.

7.3.1 Longitudinal Distributions of Particles in Jets

Consider first the parton fragmentation functions, which can be studied via the x_E distribution of particles in the unbiased away jet (7.18).

Figures 7.26,27 show x_E distributions for lower (<4 GeV/c) and higher (>3 GeV/c) $p_\perp$ of the trigger particle. The cross sections $(1/\sigma)(d\sigma/dx_E)$ scale for $p_\perp^t > 3$ GeV, the onset of scaling being evident from Fig.7.28. In the scaling region, x_E spectra from different experiments and for charged and neutral secondaries agree remarkably well (Fig.7.29) and are described by $(1/\sigma)(d\sigma/dx_E) \sim \exp(-3.7x_E)/x_E$.

The scaling violations at low $p_\perp^t$ can be understood as a consequence of a nonzero parton transverse momentum and partly as a contamination by spectator fragments /7/.

Since these effects are negligible at sufficiently large $p_\perp^t$, it seems justified to use (7.18) to predict x_E spectra for the known quark fragmentation function. For

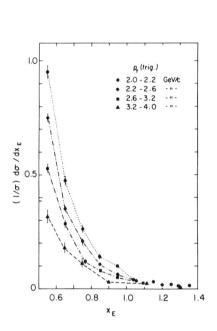
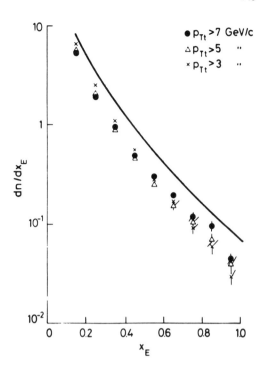

Fig.7.26. Distribution in x_E of away secondaries, $(1/\sigma)(d\sigma/dx_E)$ for four intervals of transverse momentum of the trigger. From CCHK /7/

Fig.7.27. Distribution in x_E of away-side charged secondaries for three sets of trigger $p_\perp$. The solid line is calculated from (7.23) using quark fragmentation functions with a normalization such that charged particles carry 60% of the jet momentum. From CCOR /264/

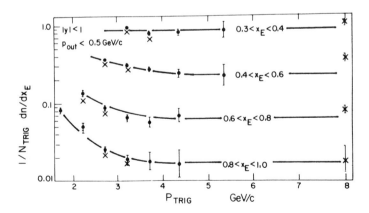

Fig.7.28. Approach to scaling of $(1/\sigma)(d\sigma/dx_E)$. The lines are drawn to guide the eye. From BFS /262/ and CCOR /264/

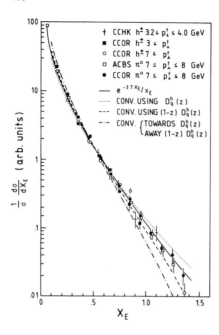

Fig.7.29. x_E distributions for charged and neutral hadrons for different ranges of trigger $p_\perp$, from CCHK /7/, CCOR /264/, ABCY /266/. Data are normalized at $x_E = 0.5$. The curves shown are calculated from (7.23) using a quark fragmentation function D_q^h for both jets (···), a "gluon" fragmentation function $D_g^h(z) = (1-z)D_q^h(z)$ for both jets (---), and D_q^h and D_g^h for the towards and away jet, respectively (-···-)

this purpose the jet cross section was taken as the measured inclusive cross section /264-266/ times the jet to single particle ratio predicted by QCD /282/. The quark fragmentation function was parameterized as $D_q(z) \sim \exp(-4.5z)/z$. Alternatively, a "gluon" fragmentation function $D_g(z) = (1-z)D_q(z)$ was used. The agreement between data and the naive calculation is surprisingly good (Fig.7.29). Data lie between the curves for quark and for gluon jets; for definite conclusions the calculation certainly is not precise enough.

It is interesting to note the similarity of the predictions using $D_q(z)$ and $D_g(z)$. The reason is that these fragmentation functions were used both for the towards and the away jet. In the "gluon" case, the distribution of fragments in z is steeper. On the other hand, the mean z of the trigger hadron decreases (7.15), and thus the momentum of the parent jets is increased for fixed trigger $p_\perp$ thereby just cancelling the additional power $(1-z)$ in $D_g(z)$. The situation changes once towards and away jets are taken to fragment in different ways; for a towards quark and an away gluon jet the prediction falls below the data at large x_E. As far as absolute rates are concerned, data typically lie slightly below the predictions (Fig.7.27). Note, however, that the normalization is essentially given by momentum conservation. Using the extrapolation $(1/\sigma)(d\sigma/dx_E) \sim \exp(-3.7x_E)/x_E$, 35% of the trigger momentum is compensated by charged particles in the away jet for the data shown in Fig.7.27!

Finally, to demonstrate what x_E scaling really means, consider Fig.7.30 where the inclusive spectrum of particles is shown together with (dn/dx_E) at $x_E = 1$. For

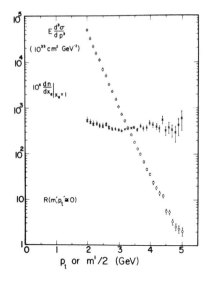

Fig.7.30. Single hadron invariant cross section and away-side multiplicity $(1/\sigma)(d\sigma/dx_E)$ for $x_E = 1$ plotted vs $p_\perp$, for p-Be interactions at 400 GeV. From E 494 /274/

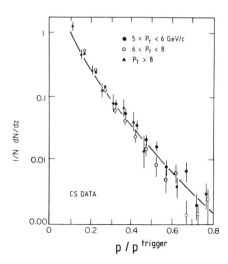

Fig.7.31. Longitudinal distribution with respect to the jet axis of charged particles associated with a high $p_\perp$ neutral trigger. Longitudinal momenta are scaled according to the trigger momentum. (From CS /265/). The solid line is calculated from (7.29) using quark fragmentation functions and is normalized arbitrarily

uncorrelated emission of toward and away hadrons, the particle density at $x_E = 1$ would drop by 5 orders of magnitude when changing the trigger $p_\perp$ from 2 to 5 GeV/c. Instead, the correlation between towards and away jet momenta keeps (dn/dx_E) constant within 20-30%!

Note in passing that at $x_E \cong 1$, the scaling limit is reached very early since the event is in a fully symmetrical configuration, and the influence of a parton $k_\perp$ nearly disappears /311/.

In analogy to (7.18) the spectra of additional particles in the towards jet are given by

$$\left(\frac{d\sigma}{d|x_E|}\right)_{towards} = \int \frac{dz}{z} E\frac{d^3\sigma}{dp^3}(\frac{p_\perp^h}{z}) D(z, |x_E|z) \qquad (7.29)$$

For a first check, the two particle fragmentation function $D(z_1, z_2)$ can be written as a product of the inclusive fragmentation functions,

$$D_q(z_1,z_2) \cong \theta(1-z_1-z_2)\exp[-4.5(z_1+z_2)]/z_1 z_2 \qquad (7.30)$$

yielding a rather good representation of the measured spectra (Fig.7.31). Although the inclusive distribution seems to scale for different trigger $p_\perp$, the fraction of momentum carried by additional particles of the towards jet decreases with increasing trigger $p_\perp$ (Fig.7.32). This is equivalent to an increase of the mean z of the trigger hadron with $p_\perp^t$ and to an increase of the jet to single particle ratio (7.12). The effect is well accounted for by (7.15) with $D_q(z)$ as chosen above. The 20% difference in normalization is explained by the azimuthal cuts used in the experiments.

As we have seen above, it is rather hard to determine the shape of fragmentation functions from x_E distributions. A more promising way is to detect as many jet fragments as possible so that the total momentum of the jet is known. Results from such attempts are shown in Figs.7.33,34: both towards and away jet fragmentation agree with each other and with lepton-induced quark jets. However, some caution is needed. Since the experiments do not manage to detect all jet fragments, large corrections are applied which in turn depend on the jet properties. It is not fully clear how unique the final, self-consistent solution really is.

Figure 7.35 shows fragmentation functions obtained from the calorimeter experiment E 260 /271/ using a jet trigger. Displayed are the measured fragmentation functions in the triggering jet and in the away jet compared to a Monte Carlo simulation based on the FF model. Whereas the z distribution in the away jet is fairly well reproduced, the authors claim to see a significant disagreement with the theory for the trigger jet; to fit the data precisely would require a very large fraction of

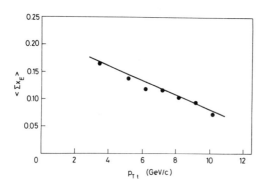

Fig.7.32. Sum of the momenta of same-side charged particles with $|\phi-\phi^t| < 60°$ scaled to the trigger momentum vs trigger $p_\perp$. From CCOR /264/. The solid curve is calculated from (7.19) using quark fragmentation functions and is renormalized by 0.8

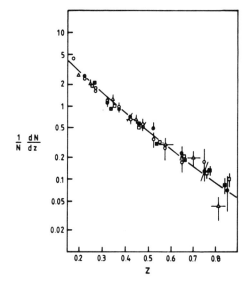

Fig.7.33. Jet fragmentation function D(z) for the towards jet (● $p_\perp^t < 6$ GeV/c, ○ $p_\perp^t > 6$ GeV/c, from CS /265/) and for the away jet (■ $p_\perp^t > 5$ GeV/c, □ $p_\perp^t > 7$ GeV/c, from CCOR /264/), compared to data from νN reactions /312/ (△) and e^+e^- annihilations /22/ (——). Distributions are normalized to unity in the interval $0.2 < z < 0.8$

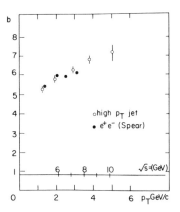

Fig.7.34. Slope parameter b of an exponential fit to the jet fragmentation function, $D(z) = \exp(-bz)$ for $0.2 \leq z \leq 0.8$, compared to data from e^+e^- annihilations /22/. From BFS /263/

gluon jets in the trigger with a fairly soft fragmentation function. The dependence of D(z) on the $p_\perp$ of the trigger is interpreted as a scale-breaking effect. It should be pointed out, however, that results on inclusive distributions in a trigger jet rely heavily on details of the Monte Carlo used to simulate the experiment (in contrast to data on ratios of jet cross sections where the major uncertainties tend to cancel), and that some problems, like the inclusion of parton transverse momentum, are not yet solved in a satisfying manner.

Figure 7.36 compares the mean charged multiplicity of jets in large $p_\perp$ events with data from e^+e^- annihilations. Although the overall agreement is not bad, especially the BSF data /263/, for low $\sqrt{s}$ the data show an increase of the mean multiplicity by slightly more than one unit. The excess can be traced to an increase of particle density by a factor 2 for $z \cong 0$ to 0.2, and the authors claim that the effect cannot be accomodated within a model with "standard" quark jets. It could be regarded as a sign for gluon jets. On the other hand, $z \cong 0.1$ corresponds to particle momenta of about 0.5 GeV/c in the cms of the two jets of large $p_\perp$, and factorization is not expected to hold for such fragments. In the spirit of the QRM

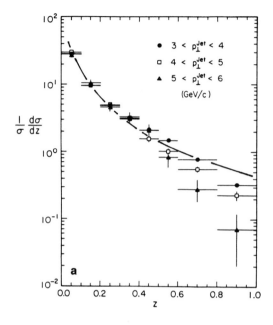
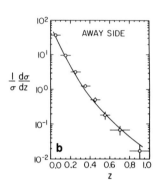

Fig.7.35a,b

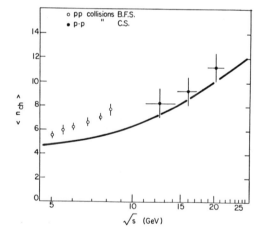

Fig.7.35. (a) Distribution of momentum fraction z scaled to the jet momentum of charged particles in the trigger jet for different ranges of the momentum of the trigger jet compared to a QCD Monte Carlo calculation (for $p_\perp$ = 4...5 GeV/c) /271/. (b) As (a), but for the away jet. Momenta are scaled to the trigger jet momentum /271/

◂ Fig.7.36. Mean multiplicity measured for large $p_\perp$ jets per jet pair together with a fit to data from e^+e^- annihilations (compare to Fig.2.2). From BFS /263/ and CS /264/

other explanations are possible. Pictorially, quarks in e^+e^- reactions produce a chain of quark pairs which "recombine" into hadrons. In large $p_\perp$ events and at small z, this chain overlaps with the primordial quark sea of the incident hadrons, and the multiplicity will be enhanced. Although such particles in principle do not "belong" uniquely to a large $p_\perp$ jet, they are close to it in rapidity and are counted by the reconstruction procedure.

7.3.2 Transverse Properties of Jets and Five-Jet Final States

As in e^+e^- annihilations, the transverse width of the jets at large $p_\perp$ is determined by two components: 1) the nonperturbative momentum smearing and 2) radiative effects as predicted by QCD. In the second case, a hard quantum is radiated by one of the active partons either before or after the basic hard scattering, yielding a five-jet final state.

In most cases, however, this new jet will not be separately observed, but will manifest itself as a broadening of the $k_\perp$ distribution of the incident partons and as an increase of the width of the jets at large $p_\perp$. Experimentally, these two effects are hardly distinguishable at present energies. A precise determination of the axis to measure the fragmentation $p_\perp$ of fast fragments suffers from the fact that a unique assignment of soft particles to the four (or more?) jets in an event is impossible. As far as fast fragments are concerned, both processes have the identical effect; they destroy the coplanarity of the event and lead to an imbalance of the momenta of the two "main" jets of large $p_\perp$.

Consider first the dependence of the fragmentation functions on the transverse momentum with respect to the jet axis, $q_\perp$. Figures 7.37,38 prove that the distribution of particles is sharply cut off in $q_\perp$, and that in large $p_\perp$ jets particles are distributed uniformly in azimuth around the jet axis. If $<q_\perp>$ is plotted as a function of the track momentum or x_E (Fig.7.39a), a seagull effect is visible. With-

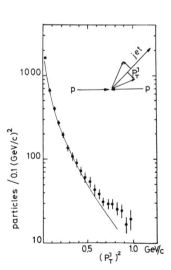

Fig.7.37. Distribution of transverse momentum with respect to the jet axis for fast secondaries in the trigger jet for $p_\perp^t \simeq 2$ GeV. Uncorrelated background is subtracted on a statistical basis. The solid line represents the $p_\perp$ distribution of secondaries in normal inelastic events, $d\sigma/dp_\perp^2 \sim \exp(-6p_\perp)$. From CCHK /310/

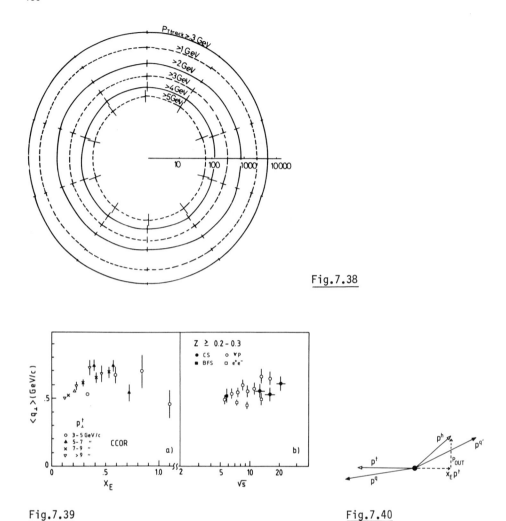

Fig.7.38. Angular distribution of jet fragments around the reconstructed axis of a jet at large $p_\perp$ for different cuts in the transverse momentum of the secondaries. Since the jets are detected at 90° in the cms, these cuts in $p_\perp$ correspond to cuts in the momentum fraction parallel to the jet axis. From CCOR /264/

Fig.7.39. Mean transverse momentum with respect to the jet axis. (a) As a function of x_E. From CCOR /264/. (b) As a function of the cms energy of the two large $p_\perp$ jets. Only hadrons at large z are taken into account. From BFS /263/ and CS /265/. As a comparison, points from lN interactions /313/ and e^+e^- annihilations /22,41/ are added

Fig.7.40. Definition of p_{out}. All momenta are projected into the transverse momentum plane. p^t, p^h, p^q and $p^{q'}$ are the momenta of the trigger particle, of a hadron in the away jet, and of the two scattered partons, respectively

in the accuracy of the measurements the values do not differ significantly from those obtained in e^+e^- annihilations (Fig.7.39b).

QCD effects can be investigated by studying the quantity $\underline{k}_\perp$, which denotes the amount of parton transverse momentum in the nucleon required to describe the effective imbalance of the transverse momenta of the two active partons after the scattering. The Q^2 or s dependence of $k_\perp$ can be measured directly in two ways. Either the component of $k_\perp$ in the scattering plane is determined by measuring the magnitude of the transverse momentum imbalance of the two jets at large $p_\perp$ which are detected in two opposite calorimeters, or one studies the component of $k_\perp$ out of the scattering plane, which gives rise to a noncollinearity of the jet transverse momenta. This is usually done by investigating the momentum component p_{out} out of the scattering plane defined by the trigger momentum and the collision axis (Fig.7.40). The mean square of p_{out} is given by

$$<p_{out}^2> = \frac{1}{2}<q_\perp^2> + x_E^2 (\frac{1}{2}<q_\perp^2> + <k_\perp^2>) \qquad (7.31)$$

where $<q_\perp^2>$ is the mean $p_\perp$ squared of jet fragments. The trigger hadron has been assumed to have $z \simeq 1$. The factor 1/2 enters since only one component of transverse momentum is used. QCD effects should manifest themselves in a rise of $<k_\perp^2>$ with Q^2 or s at fixed x_E.

Whereas for trigger transverse momenta between 2 and 4 GeV/c no variation of $<p_{out}>$ was found /7/, more recent experiments /264,266/ report a strong increase of $<p_{out}>$ with the trigger $p_\perp$ for trigger momenta between 3 and 11 GeV/c (Fig. 7.41).

Figure 7.42a summarizes the information of $<|k_\perp|>$ as derived from $<p_{out}>$ and from measurements of the momentum balance. Obviously, $<k_\perp>$ rises with the trigger $p_\perp$. In QCD, $<k_\perp>$ is expected to be proportional to a mass scale, such as $\sqrt{s}$, times a function of dimensionless variables such as $x_\perp = 2p_\perp/\sqrt{s}$ (note that $x_\perp^2 \simeq \hat{s}/s$) and α_s /320/, i.e.,

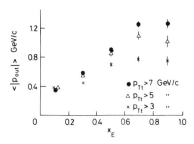

Fig.7.41. Mean value for $|p_{out}|$ as a function of x_E for different ranges of trigger $p_\perp$. From CCOR /264/

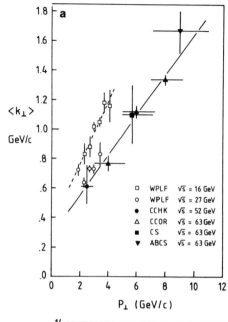

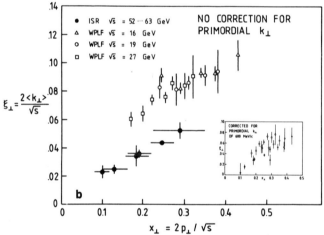

Fig.7.42. (a) Mean parton transverse momentum as determined from the WPLF /272/, CCHK /7/, CCOR /264/, CS /265/ and ABCS /266/ groups. Some of the error bars include systematic uncertainties (CCHK, CS), other refer to statistical errors only (CCOR, WPLF). (b) Dependence of the scaling quantities $\xi_\perp = 2\langle k\rangle/\sqrt{s}$ and $x_\perp = 2p_\perp/\sqrt{s}$. Small insert: same plot after subtraction of a constant primordial $k_\perp$ of 600 MeV/c from $\langle k_\perp \rangle$

$$\langle k_\perp \rangle = \sqrt{s}\, f(x_\perp, \alpha_s) \simeq \sqrt{s}\, f(x_\perp) \tag{7.32}$$

Due to the slow variation of α_s, the dependence of f on α_s can be neglected in a first approximation. Figure 7.42b shows $\xi_\perp = 2\langle k_\perp \rangle/\sqrt{s}$ as a function of $x_\perp$. Whereas data points at $\sqrt{s}$ = 16, 19, and 27 GeV scale perfectly, they disagree with values obtained at the ISR for $\sqrt{s} \simeq$ 50-60 GeV.

Part of the disagreement can be traced back to the fact that especially in the WPLF calorimeter experiment, it is not clear what the measured $\langle k_\perp \rangle$ really means since it is not known if a gluon radiated by a scattered quark is caught by the calorimeter.

Nevertheless, there is a way in which the two sets of data can be made consistent. Remember that besides the perturbative component of $k_\perp$ described by (7.32), there exists a nonperturbative $k_\perp$ inherent in the hadrons wave function. From the study of Drell-Yang processes, we know the mean primordial $k_\perp$ is about 500-600 MeV/c /314/. In fact, after subtracting quadratically a primordial $k_\perp$ of 600 MeV/c, all data approximately scale (Fig.7.42b).

To summarize so far: the fragmentation function $D(z)$ obtained for fast ($p \gg m$) fragments of the scattered partons in large $p_\perp$ events agree with the properties of quark jets measured in e^+e^- annihilations, within the experimental accuracy. Also, the transverse size of jets is similar and mainly determined by nonperturbative effects. A large asymmetry between the fragmentation of the towards and away jets, as expected in the CIM model, is excluded. On the other hand, a dominant contribution from gluons jets with a fragmentation function $D_g^h(z) = (1-z)D_q^h(z)$ seems to be incompatible with the data.

QCD effects are visible as an increase of $k_\perp$ with Q^2 and are qualitatively compatible with predictions. For definite quantitative conclusions, however, the systematic uncertainties of the experiments seem to be still too large.

7.3.3 Quantum-Number Correlations

The next step in the investigation of jets in large $p_\perp$ events is to study correlations between quantum numbers of particles emitted within one jet, or in different jets. Correlations in normal hadronic interactions are governed by the principle of "short-range order" which states that flavor quantum numbers of secondaries tend to be conserved locally /315,316/. The situation differs drastically for large $p_\perp$ events. In the hard scattering process, both confined quantum numbers (such as fractional charge or color) and nonconfined quantum numbers (such as strangeness) propagate over large distances in phase space and give rise to long range quantum number correlations.

In the remainder of Sect.7.3 we discuss correlations between the trigger flavor and hadrons in the opposite jet of large $p_\perp$. In contrast to the constituent-interchange model, QCD predicts that the flavors of the towards and the away parton emitted in proton-proton collisions (and, to a large extent, in proton-nucleus collisions) are nearly uncorrelated since in QCD subprocesses no flavors are exchanged. However, the argument does not hold for interactions where beam and target particles differ. For example, in π^-p interactions a positive towards parton always stems from the incident proton, and consequently, the away parton is a negative pion constituent. Whereas for a negative trigger parton the mean charge of away partons will be close to zero.

Let us now study correlations in proton-proton collisions. Consider first the influence of the trigger charge. Figure 7.43 shows the net charge density $(1/\sigma)(d\sigma^+/dy - d\sigma^-/dy)$ for events with a positive and negative large $p_\perp$ particle at $y^t = 0$ /262/. σ^+ and σ^- are the cross sections for production of additional positive and negative particles, respectively. It seems that the charge of the trigger particle is balanced by other particles close to it in rapidity.

A precise measure for the correlation is given by the difference of the two curves in Fig.7.43, the so called "associated charge-density balance" $\Delta q(y, y^t)$ /316/. Naively speaking, $\Delta q(y, y^t)$ is the answer to the question "which particles in a high $p_\perp$ event know about the trigger charge?" or "where do the valence constituents in the trigger particle come from?". Figure 7.44 displays $\Delta q(y, y^t)$ for events with a large $p_\perp$ hadron at $y \sim -0.9$ /317/. The peak in Δq follows the trigger rapidity; the whole distribution looks quite similar compared to the compensation of the charge of a low $p_\perp$ "trigger" particle in normal events /316/. Figure 7.45 shows Δq as a function of the azimuthal angle ϕ and the difference $y - y^t$ for two event configurations, with the two jets of large $p_\perp$ being in the same rapidity hemisphere ("back-antiback") and with the two jets in opposite hemispheres ("back-to

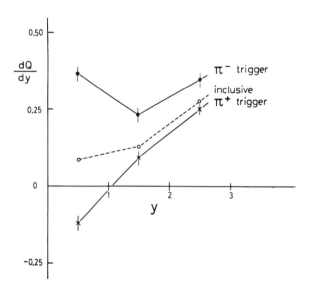

Fig.7.43. Average charge density dQ/dy for events with a large $p_\perp$ π^+ or π^- trigger at y = 0, compared to the charge density in normal inelastic events. From BFS /262/

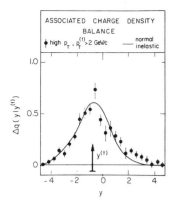

Fig.7.44. Charge-density balance $\Delta q(y, y^t)$ as a function of the rapidity y associated with a large $p_\perp$ trigger at $y^t = -0.9$. The full line shows the corresponding distribution for nondiffractive inelastic events. From CCHK /317/

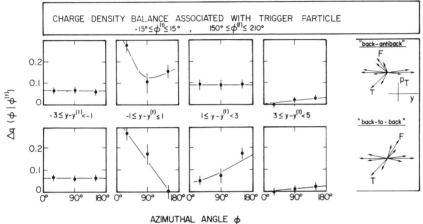

Fig.7.45. Charge-density balance $\Delta q(\phi, \phi^t)$ of particles associated with a large $p_\perp$ particle, for different intervals in rapidity relative to the trigger and for two event configurations
- large $p_\perp$ jets are in the same rapidity hemisphere ("back-antiback")
- large $p_\perp$ jets are in opposite himispheres ("back-to-back").
From CCHK /317/

back"). The rapidity of the away jet is given by the secondary with largest $p_\perp$ ("jet leader") in the away wedge $150° \leq \phi \leq 210°$. The trigger hadron is emitted at $\phi \sim 0°$, and $y \sim -0.9$.

In Fig.7.45, three components of Δq can be distinguished. Part of the charge of the trigger hadron is balanced by secondaries in the towards jet, with $\phi \sim \phi^t$ and $y \sim y^t$. A second component, peaked at $\phi \sim 180°$, follows the away jet rapidity and can be assigned to fragments of the away parton. Finally, a third component, flat in ϕ, is most naturally attributed to spectator fragments.

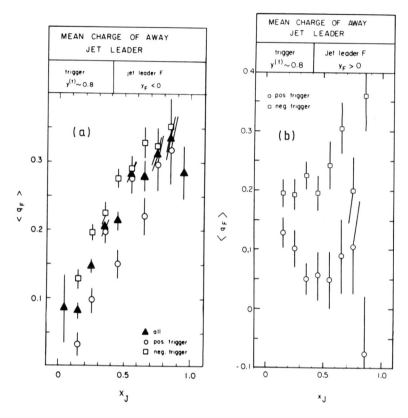

Fig.7.46a,b. Mean charge of the fastest particle in the away jet as a function of the visible momentum of the away jet scaled to the trigger $p_\perp$ for positive and negative trigger particles at $y^t \cong -0.9$. (a) Back-to-back configuration. (b) Back-antiback configuration (comp. Fig.7.51). From CCHK /259/

A contribution of away jet hadrons to Δq does not contradict QCD predictions since it may arise from soft hadrons in the region where the jets overlap, hence exchanging flavors (Fig.7.2), while the charge of the scattered parton, as measured by the charge of the away jet leader, remains unaffected. Figure 7.46a shows the mean charge of the away jet leader for "back-to-back" configurations (see Fig.7.45), in the case of positive and negative trigger particles /259/. In order to exclude cases where an undetected neutral is the leading particle in the away jet, transverse momentum balance was checked by calculating the ratio $x_J = \Sigma p_x^{tow}/\Sigma p_x^{away}$. This ratio measures the fraction of the trigger jet transverse momentum compensated by away side charged particles. In the limit $x_J \to 1$, the mean charge of the jet leader $\langle q_F \rangle$ measures the average charge of the away side parton.

In a simple quark model, a value $\langle q_F \rangle = 1/3$ is expected, since the average charge of a valence quark in a proton is 1/3. A more sophisticated calculation of valence quark scattering, based on exact structure and fragmentation functions /282/, confirms this value. The presence of gluon contributions should decrease $\langle q_F \rangle$. However, the "back-to-back" configuration selects events where $\hat{s}$, x_1 and x_2 are large, and valence quarks are preferred. In addition, the condition $x_J \to 1$ acts similarly to a trigger bias and favors quark jets as compared to QCD gluon jets.

Data are consistent with $\langle q_F \rangle \to 0.3$-$0.35$ for $x_J \to 1$. At large x_J, no significant dependence of $\langle q_F \rangle$ on the trigger charge is seen.

Figure 7.46b shows the same plot for the "back-antiback" event configuration. Here the charge of the away jet leader is correlated to the trigger charge. Although precise predictions are missing, this correlation is difficult to accommodate in the QCD model. Possible excuses are that the "back-antiback" event type corresponds to the lowest values of $\hat{s}$ for fixed $p_\perp$. Nonasymptotic exchange processes will be strongest in this configuration. Furthermore, the positive triggers contain a sizeable fraction of protons /259/, whose production mechanism was shown to differ from meson production /259/.

Flavor correlations between towards and away particles at $y \cong 0$ were investigated by the BFS group at the ISR in proton-proton collisions /262,263/ and by the experiments E 260 /271/ and E 494 /274/, E 357 /276/ at FNAL for proton-proton and proton-nucleus interactions.

Once more, it has to be emphasized that in all cases only fast away secondaries, say $x_R > 0.5$-1, provide a reasonable measure for the flavor of the away parton, and that only for $p_\perp$'s of the secondaries well above 1 GeV the background due to spectator fragments is small.

Figure 7.47 summarizes information from three experiments on the ratio of the mean number per event of fast positive and negative hadrons in the away jet for pp and pN collisions. The two experiments with trigger $p_\perp$ greater than 3 GeV/c agree that the flavor composition of the away jet does not depend on whether the trigger particle is a π^+, π^-, K^+ or p and agree with QCD predictions. The large h^+/h^- ratio obtained by E 260 for π^- triggers is probably due to a contamination from spectator fragments /271/. As far as those triggers are concerned which have no valence quarks in common with the incident hadrons, such as K^- or $\bar{p}$, the experimental situation is controversial. For example, for K^-, BFS claims significant deviations from the "standard" behavior, whereas E 494 sees no effect. More recent data from E 494 is shown in Fig.7.48. The correlation function R for two hadrons emitted with large and opposite transverse momenta is defined as $R = W_{ij}/W_i W_j$, where W_{ij} is the probability to find a flavor combination i,j among the two particle triggers and

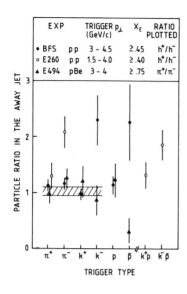

Fig.7.47. Ratio of densities of positive and negative hadrons in the away jet for different trigger particles obtained by the BFS /262,263/, E 260 /271/ and E 494 /274/ groups in proton-proton and proton-nucleus collisions

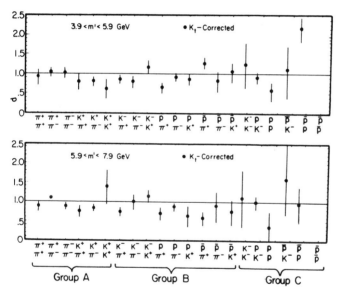

Fig.7.48. Correlation function R for two hadrons emitted with large and opposite transverse momenta for two ranges of the mass $m' = p_{\perp 1} + p_{\perp 2}$. The hadron flavors are uncorrelated if R = 1 /274/

$W_i = \sum_j W_{ij}$. It is important to note that W_i differs from the usual fully-inclusive particle ratios since triggering on two opposite hadrons virtually eliminates the effect of the parton transverse momentum smearing /274,311/! The authors consider the deviations from R = 1 as not significant /274/.

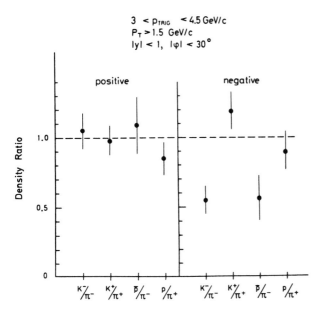

Fig.7.49. Ratio of the average number of hadrons in the away region for different combinations of trigger particles and for positive and negative secondaries. From BFS /262/

Nevertheless, one might be tempted to interpret the BFS results as an indication for CIM contributions. However, the enhancement of the h^+/h^- ratio for K^- triggers (Fig.7.47) results from a decrease of the number of negative fragments (Fig.7.49) as compared to a pion trigger, whereas for a $q\bar{q} \rightarrow K^-M^+$ process one expects the number of fast positive secondaries to rise as well. To summarize: experiments show that for well identified "nonexotic" large $p_\perp$ meson triggers, no significant correlation between the flavors of the towards and the away parton is seen. This observation once more excludes classical CIM processes as representing one of the major sources of the meson yield at large $p_\perp$. For exotic triggers (i.e., particles which do not contain a beam or target valence quark) the situation is unclear; data are sometimes contradictory and lack statistical significance.

7.4 Spectator Fragmentation

In the previous discussion of properties of the jets at large $p_\perp$, the two spectator jets at low $p_\perp$ were mainly considered as a nasty source of background. Since, however, the active partons are missing in the spectator systems, and since models like QRM or DCR predict the distribution of spectator fragments to depend crucially on its quark contents, the investigation of spectator decay offers an independent and complementary way to check the parton model concepts.

As compared to studies of spectator fragmentation in lepton-nucleon interactions, high $p_\perp$ events at ISR energies offer the advantage that the mass of the spectator systems is increased by factors of 10 as compared to the mean W in lN reactions, and therefore, it is more justified to use asymptotic concepts like DCR. Spectator fragmentation has been investigated by the CCHK/ACCDHW /259,317/ and BFS /261/ groups using the Split Field Magnet detector at the ISR. In both experiments the detector was triggered by particles of $p_\perp \simeq$ 2-3 GeV/c, which were produced in proton-proton collisions at $\sqrt{s}$ = 52 GeV. The trigger hadron was identified by Cerenkov counters.

Further results on spectator fragmentation in pp and πp collision at $\sqrt{s} \simeq$ 16-19 GeV were published by E 260 /271/. The BFS group and E 260 selected trigger particles at $y \simeq 0$, whereas the CCHK collaboration used a forward trigger at $y \simeq 2$, or $\theta \simeq 20°$. As a result, the kinematics of the two-body subprocess differs appreciably. For the central trigger with $x_\perp \simeq 0.1$, the active partons are likely to be gluons /282,286/. Because of the symmetrical configuration the partons creating the towards jet could come from each of the incident protons.

For a trigger angle of $20°$, the active parton in the proton moving in the same longitudinal direction as the trigger particle has to be in the valence quark region at $x \simeq 0.4$. Here, qq and especially qg processes are expected to be relevant. Since in QCD the forward cross sections are large compared to the probabilities for backward scattering (Fig.7.3), the towards parton will always be the quark scattered out of the proton in the same rapidity hemisphere as the trigger. The spectator fragments of this proton will be called the "accompanying spectator" as compared to the "away spectator" opposite in rapidity to the trigger.

The forward trigger condition has yet another consequence. At a fixed small angle the inclusive cross section falls more steeply in $p_\perp$ than at $\theta = 90°$, and the trigger bias pushes the mean z of the trigger hadron up to $z \simeq 0.8$-0.9 (7.15). Thus, favored fragmentation modes like $u \to \pi^+$, $d \to \pi^-$ are strongly preferred. This allows determination of the flavor of the scattered quark.

For such events with a forward large $p_\perp$ particle, the predictions of the quark parton model are evident.

- In most cases, the accompanying spectator contains two quarks, e.g., two u-quarks for a π^- trigger, and an u-quark and a d-quark for a π^+ trigger.
- Quantum numbers of the trigger particle and of the away spectator are uncorrelated.
- The away spectator is likely to contain all three incident valence quarks.

7.4.1 General Characteristics of the Spectator

Figure 7.50 shows the distribution of spectator fragments in Feynman x for a central large $p_\perp$ trigger as compared to the particle density in normal inelastic events /261/. With increasing $p_\perp^t$ of the trigger, the distributions drop more rapidly at large x. This can be understood simply as a consequence of energy conservation; the maximum energy of the spectators diminishes with increasing trigger $p_\perp$. Introducing a new scaling variable $\hat{x} = p_\parallel/(p_{beam} - ap_\perp^t)$, the BFS and E 260 collaborations found scaling for a $\cong 2$ (Fig.7.51). This value means that the amount of energy carried by the jets at high $p_\perp$ is twice the minimum amount required by momentum conservation.

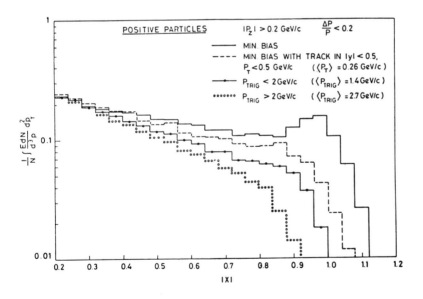

Fig.7.50. Distribution of positive spectator fragments in Feynman x for events with a large $p_\perp$ particle. Full and dashed lines refer to the corresponding distributions in inelastic and in nondiffractive proton-proton interactions. From BFS /261/

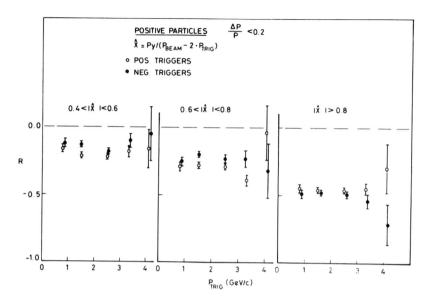

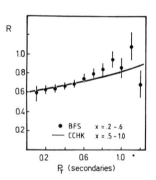

Fig.7.51. Density of positive spectator fragments in $\hat{x}$ scaled to the corresponding yields in normal inelastic interactions as a function of the trigger $p_\perp$ /261/

Fig.7.52. Ratio of particle densities in the spectator fragmentation region of large $p_\perp$ events to particle densities observed in normal inelastic interactions as a function of the particles $p_\perp$. From BFS /261/ and CCHK /258/

This may seem embarrassing. However, Monte Carlo studies prove that for a = 2, the variable $\hat{x}$ is a good approximation to the "correct" variable x', defined in (7.20), when averaged over the away jet rapidities. So, Feynman scaling seems to hold for spectator fragmentation.

Let us now turn to transverse momentum distributions in spectator jets. Figure 7.52 shows the ratio R of particle densities in spectator jets and in normal inelastic events as a function of $p_\perp$ with respect to the collision axis. Once more the transversely cut-off distribution of momenta proves to be an universal feature of jets, the mean transverse momenta in the two classes of events deviate by less than 10%. Assuming that the basic fragmentation mechanisms are identical, the small difference in $<p_\perp>$ can be explained in different ways:

- Contamination by particles from the jets at large $p_\perp$. Since the rapidity of such particles is essentially restricted to $|y| \lesssim 2-3$, the effect should be negligible for $|x| \gtrsim (2m_\perp/\sqrt{s}) \sinh(y_{max}) \cong 0.3-0.4$, for $p_\perp \lesssim 1$ GeV,
- gluon bremsstrahlung of the active quark before the scattering,
- the spectator axis of flight deviates from the collision axis since the active parton recoils against the spectator.

Such a recoil has in fact been observed /258/ as a pronounced asymmetry in the azimuthal distribution of fast spectator protons in high $p_\perp$ events (Fig.7.53). Figure 7.54a shows the mean transverse momentum component opposite to the trigger $p_\perp$ as a function of the trigger $p_\perp$ for a central trigger. The amount of recoil carried saturates for higher $p_\perp$ of the trigger. Such a saturation is expected for the primordial component of parton $k_\perp$. The comparison of the recoil component and of the effective $k_\perp$ measured by the imbalance of momenta of the large $p_\perp$ jets (Fig.7.42) offers the possibility to separate soft gluon contributions (primordial $k_\perp$?) from hard gluon effects (QCD). Only soft gluons are expected to be reabsorbed by the spectators; hard gluons should materialize as separate jets. Present data indicate a different dependence on the trigger $p_\perp$ for the two components, but for definite conclusions, more precise data on the recoil momentum for larger $p_\perp^t$ are needed.

Concerning the third possibility listed above, the BFS group claims that the increase of the mean $p_\perp$ of detector fragments cannot be fully explained by a momentum smearing due to the recoil.

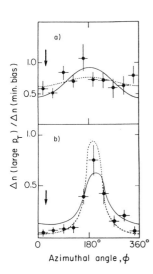

Fig.7.53a,b. Azimuthal distribution of fast positive particles (a) in the away spectator in large $p_\perp$ events, normalized to the particle density in normal inelastic events. Full and dotted lines refer to parton model calculations including parton $k_\perp$. From CCHK /7,258/, (b) in the accompanying spectator

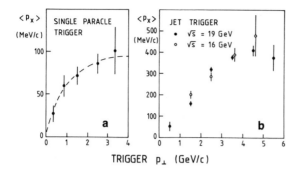

Fig.7.54a,b. Mean component of transverse momentum opposite to the trigger $p_\perp$ as a function of the trigger $p_\perp$ (a) per secondary, from BFS /261/ at $\sqrt{s}$ = 52 GeV, (b) per spectator jet, from E 260 /271/ at $\sqrt{s}$ = 16-19 GeV

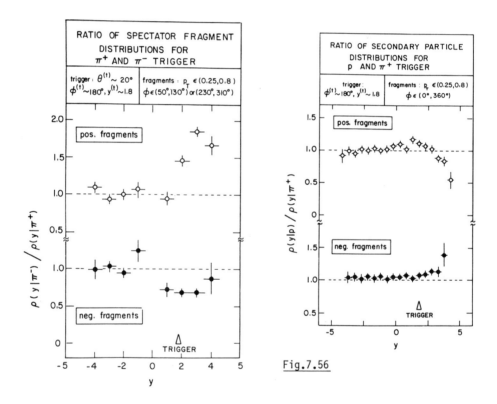

Fig.7.55. Ratio of the distributions of spectator fragments for a π^+ trigger $[(1/\sigma)(d\sigma/dy)_{\pi^+ \text{ trigger}} \equiv \rho(y|\pi^+)]$ and for a π^- trigger $[\rho(y|\pi^-)]$ as a function of rapidity. The region $y > 0$ is populated by fragments from the accompanying spectator. From CCHK /259/

Fig.7.56. As Fig.7.63, but for proton and π^+ trigger. From CCHK /259/

7.4.2 Quantum-Number Correlations

As already discussed, the quark-parton model predicts strong correlations between trigger and spectator quantum numbers. Figure 7.55 shows the ratio of rapidity distributions of positive secondary particles for a π^+ trigger to those with a π^- trigger and similarly for negative secondaries. The trigger particle at $y^t \simeq 2$ has a mean $p_\perp$ of 2.5 GeV/c. Similarly, the corresponding ratios for p/π^+ triggers and for K^-/π^- triggers are given in Figs.7.56,58 (note that the "K^- triggers" contain a certain , ~30%, contamination of large $p_\perp$ antiprotons and that the "p trigger" includes some K^+, ~20% /259/. A clear correlation is seen between the nature of the trigger particle and charged particles in the accompanying spectator jet; for $y > 2$, where spectator fragments dominate, the ratios deviate markedly from unity.

The size of the effect is as expected: the accompanying spectator in π^+ triggered events, which is a ud-quark system, contains less positive fragments than the uu-spectator in π^- triggers. The different quark composition of the accompanying spectator is once more demonstrated in Fig.7.57, where the $p\pi^+$ mass spectra is shown

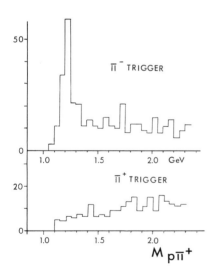

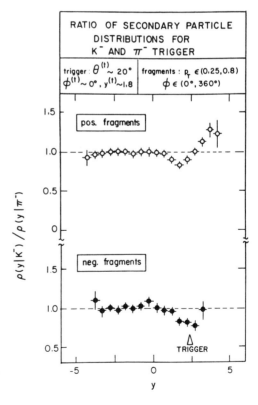

Fig.7.57. $p\pi^+$ invariant mass distributions in the accompanying spectator of events triggered on a forward π^+ or π^- meson at large $p_\perp$. From ACCDHW /318/

Fig.7.58. As Fig.7.63, but for K^- and π^- triggers. From CCHK /259/

for fast fragments. In π^- triggered events (uu-spectator) a clear Δ^{++} (uuu) signal is visible, which disappears completely for π^+ triggers (ud-spectator).

7.4.3 Spectator Fragmentation Functions

For a forward π^+ and π^- trigger particle, the quark composition of the accompanying spectator are assumed to be ud and uu, respectively. Using dimensional counting rules or the quark recombination model, the scaling distribution of fast fragments can be predicted and compared with experiment. Vice versa, one can try to identify the scattering mechanism by inspecting the quark composition of the spectator, an application which is particularly interesting for the proton and K^- triggers.

As a scaling variable, we use x' as defined by (7.20) and make the following approximations for y, y' and $m_\perp$:

$y = y^t$ (trigger rapidity),
$y' = 0$ (experimentally: $\langle y'\rangle = 0.05$),
$m_\perp = 1.1\, p_\perp^t$

The effects of these approximations have been simulated, showing that an initial distribution $Ed^3\sigma/dp^3 \sim (1-x)^\beta$ can be regained with an error in β of less than ~ 0.5. For ratios of cross sections, these errors tend to cancel.

To compare with data, we use the QRM as defined by (6.24). Numerical values for meson production are taken from /201/. Predictions for ratios of particle densities are absolutely normalized.

In addition predictions from counting rules are quoted, calculated via $x'd\dot\sigma^h/dx' \sim (1-x')^{2n-1}$, where n is the minimum number of quarks left over after the hadron h is emitted (Sects.6.5,6). The normalization is arbitrary. In apparent contradiction with experiments /319/, such a counting rule predicts the distribution of π^+ and π^- from proton fragmentation to be identical. To correct for this effect, we use an ad hoc modification /192,259/ of the counting rule recipe: the predicted power is increased by one unit whenever the fragment contains a valence d-quark arising out of one of the primary protons.

The same modification has been applied to the DCR predictions assuming pointlike particle creation (6.37). DCR predictions are summarized in Table 7.2.

For the K^- trigger, three possibilities have been considered:

- scattering of a valence quark followed by an unfavored fragmentation into a K^-;
- scattering of a $\bar{s}$ sea quark,
- scattering of a gluon.

Table 7.2. Dimensional counting rule predictions for spectator fragmentation spectra

Trigger type	Hard scattering process	Spectator contents	Fragment charge and type		Predicted spectrum "standard" DCR	Predicted spectrum "pointlike" DCR
π^+	$qq \to qq$, $qg \to qg$	ud	+	p	$(1-x)$	$(1-x)$
			−	π^-	$(1-x)^4$	$(1-x)^4$
π^-	$qq \to qq$, $qg \to qg$	uu	+	p	$(1-x)$	$(1-x)$
			−	π^-	$(1-x)^7$	$(1-x)^5$
k^-	$qq \to qq$, $qg \to qg$	uu, ud	+	p	$(1-x)$	$(1-x)$
			−	π^-	$(1-x)^4$	$(1-x)^4$
	$q\bar{q} \to q\bar{q}$, $\bar{q}g \to \bar{q}g$	uud$\bar{s}$	+	p	$(1-x)$	$(1-x)$
			−	π^-	$(1-x)^8$	$(1-x)^{4-8}$ b)
	$gg \to gg$, $gq \to gq$	uud	+	p	? a)	? a)
			−	π^-	$(1-x)^6$	$(1-x)^4$

(a) The straight forward prediction $(1-x)^{-1}$ is obviously not adequate.)
(b) Depending on whether the sea quark belongs to the primordial wave function or not.)

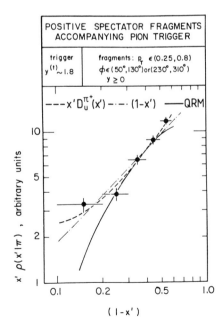

Fig.7.59. Distribution in x' of positive fragments in the spectator accompanying large $p_\perp$ pions compared to favored quark fragmentation functions $D_u^{\pi^+}$, the counting rule prediction $(1-x')$, and the prediction of the QRM, for a diquark-quark recombination. From CCHK /259/

Figure 7.59 shows the distribution of positive fragments in the spectator accompanying a pion trigger. The DCR prediction fits the data reasonably well, whereas the QRM distribution drops a little too fast at large x, an effect which had been observed already in 1N interactions (Fig.6.30). As a less model dependent comparison, the quark fragmentation function $D_u^{\pi^+}$ is shown, assuming that the fragmentation of a quark into a meson will be similar to the fragmentation of a diquark into a baryon.

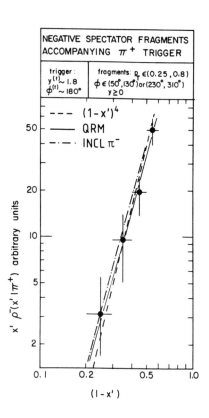

Fig.7.60. Spectrum of negative particles in the spectator accompanying a π^+ trigger compared to model predictions and to the fully inclusive π^- spectrum. From CCHK /259/

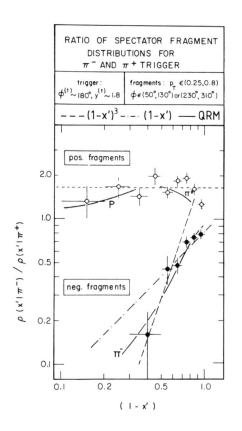

Fig.7.61. Ratios of distribution of spectator fragments in events with a π^- and a π^+ trigger. The QRM prediction for negative fragments refers to pions; for positive fragments, pions were assumed to dominate at small x' and protons at large x'. From CCHK /259/

The spectrum of negative particles in the spectator accompanying a π^+ trigger is given in Fig.7.60. The shape of the spectrum coincides with the fully inclusive pion distribution in proton-proton collisions /319/ and is reproduced by both QRM and DCR models.

To demonstrate the influence of the trigger charge the ratios of x' distributions for secondaries in the spectator jet accompanying a π^- trigger and a π^+ trigger are shown in Fig.7.61 for both positive and negative secondaries. As expected, there are less fast negative fragments in π^- triggered events than for π^+ triggers, proving that in the first case the valence d-quark has been struck. The model calculations included in Fig.7.61 are straightforward as far as negative secondaries are concerned; the QRM model describes data well even at low x'. The prediction based on "standard" DCR seems to be too steep at low x', but below $x' \simeq 0.3-0.4$ the model is not expected to hold anyway. The QRM values for positive particles include three components. At low x', the ratio is dominated by π^+ production. At large x' the ratio is determined by the proton yields from

i)   ud → p   + X for trigger π^+
ii)   uu → p   + X for trigger π^-
iii) uu → Δ^{++} + X for trigger π^-
           → pπ^+

Based on the direct term i) and ii) alone, one expects the ratio plotted in Fig.7.61 to increase with increasing x' since in the average the uu-system has a larger momentum than the ud-spectator. Contributions from iii), however, increase the ratio to values larger than 1 in agreement with experiment. Note that the prediction is not sensitive to the assumption that all baryons are produced in their ground states.

The corresponding ratios of particles produced in association with a K^- and π^- trigger are shown in Fig.7.62. In the K^- events, positive spectator fragments are enhanced and the density of negative particles is reduced. This result is compatible with the prediction of standard DCR for scattering of strange quarks, or with pointlike DCR for strange quark or gluon scattering, taking into account that the ratio for negative fragments is probably still compatible with ~const for $x \to 1$. It is, however, hard to accommodate these observations in the framework of a recombination model. Since the K^- contains no d-quarks, it seems impossible to imagine any scattering mechanism which picks up the valence d-quark with the same efficiency than for π^- triggers. Without such a mechanism, the ratio shown in Fig. 7.62 is predicted to increase with x', for negative secondaries!

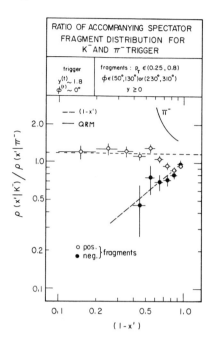

Fig.7.62. As Fig.7.61 but for K^- and π^- triggers

Another example where new information can be derived from the analysis of spectator fragmentation is the production of forward large $p_\perp$ baryons in proton-proton collisions. There are strong indications that high $p_\perp$ protons result from the scattering of a quasi-bound diquark system out of one of the incident protons. A detailed discussion of this point is found in /259/.

Let me summarize so far: using the dimensional counting rules, the shape of particle spectra in the accompanying spectator can be described. The quark recombination model using a Kuti-Weisskopf matrix element fits the distribution of spectator mesons in events with pion triggers quite well, but fails for "exotic" large $p_\perp$ particles like K^-. There are two possible reasons for this failure.

a) The Kuti-Weisskopf model for the proton wave function makes a somewhat artificial distinction between valence quark and sea quark matrix elements. It may happen that although the model accounts for correlations between valence quarks, it underestimates the correlation between a valence quark and a fast sea quark.

b) In the QRM the fragments reflect the instantaneous distributions of quarks at the moment of the scattering. There, the spectator is far from its equilibrium. After scattering a quark at $x \simeq 0.3$, 70% of the spectator momentum turns out to be carried by gluons. Assuming that this system regains its equilibrium before the fragmentation starts, the qualitative agreement with experiments can be improved.

7.4.4 Particle Correlations in Spectator Jets

The short-range correlations observed between hadrons emitted in normal inelastic hadron reactions are usually described in terms of a cluster model /64,316,321/, assuming that in a first stage of the interaction, hadronic clusters with masses of about 1-1.5 GeV are produced. There is some evidence that a "cluster" is a synonym for a sum over the known vector and tensor meson resonances /322/.

The mean decay multiplicity and the width of the correlation induced by cluster decay are characteristic and energy independent features of low $p_\perp$ interactions.

To investigate such short-range rapidity correlations among spectator fragments, the perturbation by particles from the two jets at large $p_\perp$ has to be minimized. To achieve this goal the CCHK group /317/ studied events where both the towards and the away jet are in the same rapidity hemisphere at rapidities $y < -0.7$. The jet rapidities are taken as the rapidity of the trigger particle and as the rapidity of the fragment with largest $p_\perp$ in the away jet region.

Correlations were studied among spectator fragments at positive rapidities. Figure 7.63 shows the two particle density correlation

$$C'(y_1|y_2) = \rho(y_1|y_2) - \rho(y_1)$$

for $y_2 \cong 2$ and $y_2 \cong 4$. The correlation function $C'(y_1|y_2)$ describes the change in particle density at y_1 for events which have a particle at y_2 as compared to the inclusive rapidity distribution. C' is closely related to the correlation function C defined by (4.47) /316/. From Fig.7.63 a strong short-range correlation is evident. Both size and width of the correlation essentially agree with the corresponding distributions obtained in ordinary nondiffractive proton-proton collisions at the same cms energy /316/. In Fig.7.64, the charge balance $\Delta q(y_1,y_2)$ is plotted as a function of the rapidity y_1. $\Delta q(y_1,y_2)$ (Fig.7.44) measures where in rapidity the charge of a particle selected at y_2 is balanced. Charge is seen to be conserved locally in phase space, like in normal nondiffractive hadron reactions /316/.

The similarity of correlations in inelastic events at low $p_\perp$ and in spectator fragmentation at large $p_\perp$ suggest that the same mechanism of fragmentation or confinement occurs, supporting the hypothesis of jet universality in its most general sense. Moreover, the observed correlations are in qualitative agreement with the preconfinement concept in QCD which states that during the evolution of the rapidity plateau color singlet clusters with masses of the order of 1 GeV are created. From parton diagrams describing jet formation via successive branching (Fig.5.6), one learns that these states tend to be flavor neutral. Assuming that such clusters

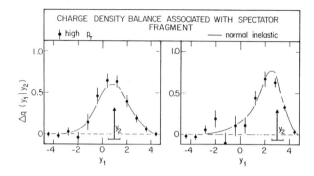

Fig.7.63. Two-particle density correlation $C'(y_1|y_2)$ for spectator fragments in large $p_\perp$ events for two rapidities $y_2 = 2$ and $y_2 = 4$. The full lines show the corresponding distributions in normal inelastic events. From CCHK /317/

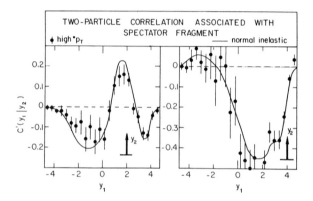

Fig.7.64. Associated charge-density balance $\Delta q(y_1|y_2)$ for spectator fragments as a function of y_1 for $y_2 = 2$ and $y_2 = 4$. Full lines show the corresponding distributions obtained in normal inelastic events. From CCHK /317/

decay without a large reshuffling of quark lines, one immediately predicts both size and width of the observed correlations and explains the observed local conservation of charge.

7.5 Summary

The distribution in phase space of secondaries in hadron-hadron interactions with large $p_\perp$ particles is consistent with the four-jet structure characterizing a basic two-body parton-parton scattering. The main features of both inclusive spectra and particle ratios at large $p_\perp$ are described in the QCD model where quark-quark, quark-gluon and gluon-gluon subprocesses contribute. Especially at medium $p_\perp \cong 2\text{-}5$ GeV/c, scale-breaking effects and corrections due to parton $k_\perp$ have to be taken into account.

Significant contributions from constituent interchange processes, like $q\bar{q} \to M\bar{M}$, or $qM \to q'M'$ are excluded.

The properties of the jets at large $p_\perp$ agree with those of quark jets observed in e^+e^- annihilations, except for a possible increase of the plateau height. There is no clear evidence for the presence of gluon jets with fragmentation functions markedly different from quark fragmentation. However, based on the experience gained from the study of T decays, such a difference is not expected to show up at jet momenta below 5 to 10 GeV.

In agreement with QCD predictions, the quantum numbers of partons at large $p_\perp$ are essentially uncorrelated, proving that no flavors are exchanged in the subprocess. Only for production of "exotic" particles which cannot result from favored fragmentation of a scattered valence quark, certain correlations are observed. However, the experimental situation is still ambiguous.

A further test of the QCD model was obtained by studying spectator fragmentation. It is shown that in proton-proton reactions with a forward large $p_\perp$ π^+ and π^-, an u- and d-quark, respectively, is scattered out of the proton moving in the same longitudinal direction as the pion, proving further that the underlying parton-parton cross section is strongly peaked forward.

A comparison of $p_\perp$ distributions and two particle correlations in the spectator region with those obtained in normal inelastic events proves that similar mechanisms of fragmentation act in both cases.

8. Hadron-Hadron Interactions at Low $p_\perp$

Motivated by the success of the quark-parton model in the description of deep inelastic processes, one is tempted to apply this picture to normal inelastic hadron interactions as well.

Many of the general aspects of low $p_\perp$ hadron interactions have been referenced already in Chaps.2 and 7.

- The secondaries form jets around the collision axis.
- Similar to the jets observed in lepton induced reactions, these jets consist of a fragmentation region carrying the quantum numbers of the incident particle and of a plateau region. The retention of quantum numbers like charge is illustrated in Fig.8.1, where particle density and charge density distributions per event are shown as a function of the rapidity y, for pp reactions at $\sqrt{s}$ = 52 GeV /316/.
- The s dependence of particle spectra is described by the concept of limiting fragmentation /2,323/. Inclusive spectra in the fragmentation region scale in Feynman x, respectively in x_E = $2E/\sqrt{s}$. With increasing s, the length of the plateau increases, but its other characteristics essentially stay unchanged (Fig.8.2).
- Inclusive fragmentation spectra factorize in a sense that at high energies, beam and target fragmentation depend only on the type of beam and target particle. This is visualized in Fig.8.3.
- Transverse momenta with respect to the collision axis are limited. Mean momenta are of the same size as the nonperturbative component of $p_\perp$ in reactions at large momentum transfers.

All this is precisely what is expected in parton models where fragmentation spectra are related to the distribution of valence quarks in the incident particles /13-15/. In the following section, we shall concentrate on the main characteristic of parton processes, the x distribution of fragments.

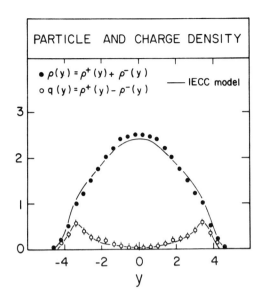

Fig.8.1. Particle and charge density in high-multiplicity proton-proton reactions at $\sqrt{s}$ = 52 GeV as a function of the rapidity y. From CCHK /316/

8.1 Longitudinal Fragmentation Spectra

To apply models like DCR or QRM to reactions at low $p_\perp$, one has to identify the basic interaction which gives rise to the fragmentation. In deep inelastic scattering the process initiating the fragmentation is well known: a large momentum transfer leads to a separation of color carriers in space.

The existence of a rapidity plateau connecting the fragmentation regions of low $p_\perp$ events indicates that in the initial interaction "something" is exchanged between the incident hadrons which gives rise to a long-range force. The most natural candidate for this "something" is color. Within the general concept of QCD there are two mechanisms which could provide a color exchange.

- color-octet, vector-gluon exchange (Fig.8.4b) /325-327/,
- color-triplet quark exchange, respectively quark-antiquark annihilation (Fig.8.4a) /4,15,73/.

The gluon exchange mechanism was the starting point for the Pomeron model of LOW /325/ and NUSSINOV /326/ and has the advantage of automatically generating a constant high-energy cross section.

The quark exchange or annihilation mechanism is the QCD equivalent of Feynmans wee-parton exchange /4/.

The quark exchange model yields total cross sections rising as ln(S) for sea-quark-sea-quark interactions and constant cross sections for valence-quark-sea-quark

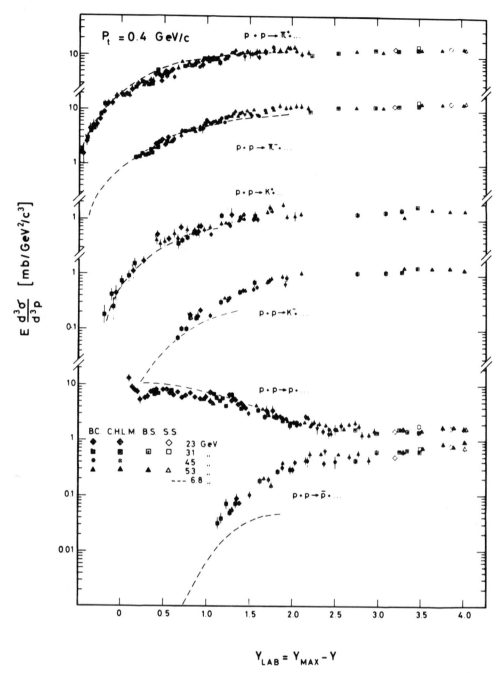

Fig.8.2. Inclusive distributions of particles produced in proton-proton interactions at $\sqrt{s}$ = 6.8 to 53 GeV plotted as a function of $y_{MAX}-y \cong -\ln(x)$ at fixed $p_\perp$. /324/

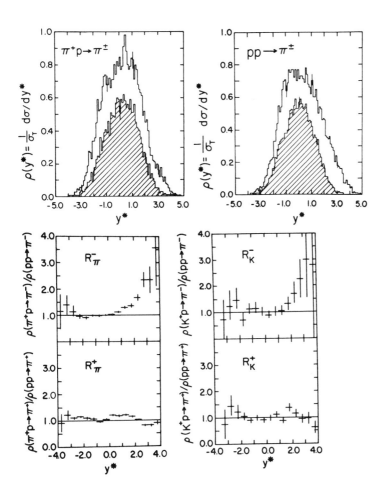

Fig.8.3. (a) Particle density per event $(1/\sigma)(d\sigma/dy)$ as a function of the cms rapidity y for $\pi^+p \to \pi^+$, π^- (shaded) and pp $\to \pi^+$, π^- at $\sqrt{s} \cong 14$ GeV. (b) Ratios of the positive and negative pion densities for $\pi^+p \to \pi+X/pp \to \pi+X$ ($R_\pi^\pm$) and for $K^+p \to \pi+X/pp \to \pi+X$ ($R_K^\pm$). Particle densities in the target fragmentation regions (y < 0) are independent of the type of the beam particle /328/

reactions /193/. Processes such as valence-quark-valence-quark fusion (e.g., in πN reactions) give a negligible contribution to the total high-energy cross section since the valence wave function of beam and target barely overlap.

A possibility to distinguish between the quark and gluon exchange mechanisms has been proposed by BRODSKY and GUNION /193/. It is based on the ansatz that particle spectra at large x are determined by those quark diagrams containing the minimum number of quark lines. According to the DCR, for example, each additional spectator line damps the x distributions of fragments by at least (1-x) (see Sect.6.6). Consider

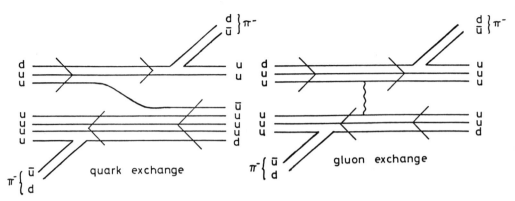

Fig.8.4. Proton-proton interactions initiated by (a) quark exchange, (b) gluon exchange. The diagrams show the minimum graphs describing the production of two fast pions at opposite rapidities /193/

now the reaction pp → π+π+X with two fast pions at opposite rapidities. The relevant minimal diagrams are displayed in Fig.8.4. One recognizes that for gluon exchange the production mechanisms are identical for both fragmentation regions; the pion momenta are not correlated. For a quark exchange to occur, one of the protons has to be in a higher Fock state containing sea quarks. In the framework of counting rules, the momentum sharing among at least five quarks results in a damping of quark structure functions and consequently of the pion spectra at large x. The production rates for fast mesons in opposite fragmentation regions are expected to show a pronounced anticorrelation /193/.

The corresponding predictions based on the recombination model are less evident and depend on the choice of the multiquark wave function. Nevertheless, it is evident that in a state as described by the upper half of Fig.8.4a where one of the valence quarks is forced to be at low x, the mean momentum of the remaining valence quarks is larger than in a state as given by the lower part of Fig.8.4a.

Correlations between two fast pions have been studied at the ISR for $\pi^+\pi^+, \pi^+\pi^-$, and $\pi^-\pi^-$ combinations /329/. Figure 8.5 shows the correlation function

$$R = \sigma_{tot}\left(\frac{d^6\sigma}{d^3\underline{p}_1 d^3\underline{p}_2}\right)\bigg/\left(\frac{d^3\sigma}{d^3\underline{p}_1}\frac{d^3\sigma}{d^3\underline{p}_2}\right) \tag{8.1}$$

as a function of the Feynman x of the pion with momentum $\underline{p}_2$ for different momenta of pion 1. Full and dashed-dotted lines indicate DCR and QRM predictions for quark exchange, respectively. Dashed lines refer to uncorrelated emission after a gluon exchange. Obviously, any substantial contribution due to quark exchange is ruled out.

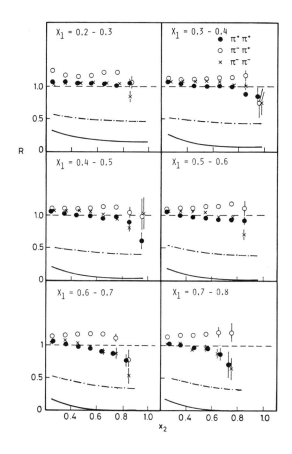

Fig.8.5. Correlation coefficient R for emission of fast opposite pions plotted vs x_2 at fixed x_1 for $\pi^+\pi^+$, $\pi^+\pi^-$, and $\pi^-\pi^-$, in pp interactions at $\sqrt{s}$ = 63 GeV /329/. The full line shows a prediction for quark exchange based on dimensional counting rules (Sect.6.6), the dashed-dotted line refers to quark exchange and quark recombination, using a Kuti-Weisskopf matrix element (Sect.6.5). The dashed line corresponds to gluon exchange

We are now prepared to predict single inclusive fragmentation spectra using the recombination model or dimensional counting rules.

For a quantitative comparison, collective excitation and dissociation of the incident hadrons also have to be taken into account besides the "quasi-perturbative" processes given by the QRM /193,197/. Conventionally, such processes are described in the triple-Regge scheme. One obtains for $A+B \to C+X$

$$E\frac{d^3\sigma}{dp^3} \simeq (1-x_c)^{1-2\alpha_{A\bar{C}}(t_{AC})}$$
$$\text{for } x_c \to 1 \tag{8.2}$$

where $\alpha_{A\bar{C}}(t)$ is the leading Regge trajectory identified in the exclusive reaction A+H → C+H'. The x and $p_\perp$ dependence of σ given by (8.2) allows to identify triple-Regge and parton contributions experimentally. For triple-Regge terms, the slope in $p_\perp$ of $Ed^3\sigma/dp^3$ increases with increasing x, whereas for incoherent parton processes one expects factorization of the type

$$E\frac{d^3\sigma}{dp_c^3} \sim (1-x_c)^n g(p_\perp) \tag{8.3}$$

Triple-Regge contributions turn out to be important /193,197/ for reactions such as,

$$\begin{aligned} A+B &\to A'+X \\ p+p &\to p+X \\ \pi+p &\to \rho+X \end{aligned} \tag{8.4}$$

where A and A' contain the same quarks. In quantum number exchange reactions, Regge terms are sizeable only at large $x \gtrsim 0.9$ /197/. Thus, reactions of the type (8.4) are omitted in the following discussion, and the x range used is restricted to $0.3 \lesssim x \lesssim 0.8$. The lower x cut excludes central production mechanisms and reduces effects due to resonance decays.

The data quoted in the following discussion come from experiments performed at cms energies between 14 and 63 GeV /319,330-333/. To compare with DCR predictions, n and $g(p_\perp)$ are used as determined from fits of (8.3) to the data. At FNAL energies, it is preferable to use the scaling variable $x_R = 2E/\sqrt{s}$ instead of Feynman x since the scaling limit was shown /334/ to be approached faster in x_R. For the reactions considered, no significant deviations from factorization in x and $p_\perp$ (8.3) were seen /319,333/.

The data available are summarized in Fig.8.6a-c. Since in most models the exponent n essentially depends only on the type of beam particle (meson or baryon), on the type of the beam fragment, and the number of common valence quarks, data like

$$\begin{aligned} \pi^+ N &\to K^+ + X \\ \pi^- N &\to K^- + X \\ K^+ N &\to \pi^+ + X \\ K^- N &\to \pi^- + X \end{aligned}$$

are averaged over. The error bars given indicate the statistical error, resp., the rms spread of n when averaged over several, theoretically identical processes, whatever is larger. The motivation for the second choice is that the spread in n mea-

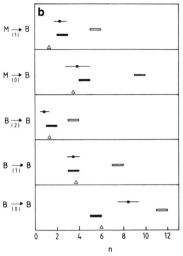
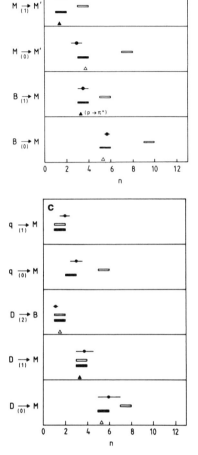

Fig.8.6a-c. The exponent n describing particle production in the fragmentation region $0.3 \lesssim X \lesssim 0.8$ /319,330-333/. (a) Production of a meson M̃ in the fragmentation region of a meson M or a baryon B. (b) Production of a baryon B. (c) Production of quark and spectator fragments in reactions involving large momentum transfers ("D" denotes a diquark system). m is the number of common valence quarks of initial hadron and fragment. Predictions of QRM are shown as triangles, "standard" and "pointlike" DCR are shown as open and closed bars, respectively

sures the influence of finite mass effects, resonance decays, and spillover from the target fragmentation region, and thus indicates the inherent limitations of such simplifying models.

The nomenclature is as follows:

$$H \underset{(m)}{\to} H'$$

where H is an incident hadron, H' is a particle out of the corresponding fragmentation region, and m is the number of quarks common to H and H'.

Figure 8.6a presents values of n for meson production by meson and baryon beams, Fig.8.6b displays the corresponding values for baryon production. Finally, the results on spectator fragmentation in high $p_\perp$ reactions are shown in Fig.8.6c.

Theoretical predictions are given for three different models.

a) Quark recombination (Sect.6.5) enables very detailed predictions for processes where the fragment contains one or more valence quarks of the incident hadron. However, the knowledge of the corresponding valence structure function is required. This condition restricts the predictive power of the model to processes where fast mesons are produced. QRM predictions are shown in more detail in Fig. 8.7 for $pp \to \pi^++X$ and $pp \to \pi^-+X$ /197/. Here, the maximum amount of information is available both theoretically and experimentally. The shape of particle spectra is extremely well reproduced, and questions concerning the absolute normalization have been discussed in Sect.6.5. In Fig.8.6, definite predictions of the QRM are shown as black triangles, whereas values relying on the specific choice of the Kuti-Weisskopf structure function are given by open triangles.

b) Counting rules for pointlike emission [Sect.6.6 (6.47)] is based on the same philosophy as the QRM: the distribution of fragments reflects the instantaneous distribution of quarks at the moment of the interaction. The predicted power n may be increased by one unit due to spin or isospin effects, e.g., for $pp \to \pi^-+X$ as compared to $pp \to \pi^++X$. The full bars shown in Fig.8.6 include this uncertainty.

c) "Standard" counting rules [Sect.6.6, (6.40)]. Here it is assumed that an equilibrium state containing all final state quarks preceeds particle emission (open bars in Fig.8.6). ·

Figure 8.6 proves that these standard counting rules clearly disagree with data from $p_\perp$ hadronic interactions. Both "pointlike" counting rules and the quark recombination model are in reasonable agreement with low $p_\perp$ data and with results on spectator fragmentation in events with large momentum transfers, with the following exceptions:

- the distributions of antibaryons in hadronic collisions is predicted too flat;
- the distribution of negative fragments in the accompanying spectator in events with a large $p_\perp$ K^- is predicted too flat (Fig.7.62).

As already noticed, part of these discrepancies could be cured by choosing a more complicated proton wave function which includes dynamical correlations between valence and sea quarks; for definite conclusions more precise data are needed.

At this point the question arises if these "soft" hadronization models can be applied to describe the fragmentation of quark or gluon jets at large Q^2, e.g., if "hard" and "soft" jets can be understood in a common framework.

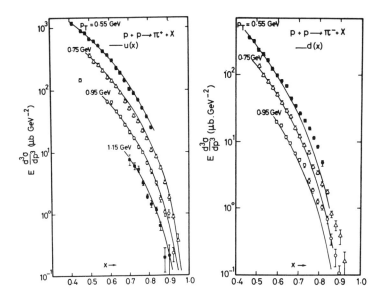

Fig.8.7. Inclusive $\pi^{\pm}$ cross sections at ISR energies compared to predictions based on the QRM. Deviations for $x \gtrsim 0.9$ result from tripple-Regge contributions /197/

Figure 8.6c shows the predictions of "pointlike" counting rules for the favored and unfavored fragmentation function for quark jets. The predictions $(1-x)$ and $(1-x)^2$ are slightly too flat in x compared to the measured spectra for $0.3 \lesssim x \lesssim 0.8$ which is the same range of x as used in the discussion of spectator fragments.

An explanation for this tendency may be given by scale-breaking effects: in the evolution of a quark jet, the formation of final state hadrons will occur at some scale $Q_0^2 \cong 1$ GeV$^2 \ll Q^2$. The fact that hadron fragmentation is described by the QRM using the structure functions obtained in deep inelastic reactions indicates

$$Q^2 \gg Q_0^2 > \Lambda^2 \tag{8.5}$$

where Λ is the QCD scale parameter. In the fragmentation from Q^2 to Q_0^2 partons are generated which increase the number of spectator fields as compared to the minimum number [see (6.36)]:

$$zD(z) \cong (1-z)^{n_{eff}(Q^2)-1}$$

Recently quantification of these ideas within the framework of the QRM has been tried /335/. QCD (5.27) is used to describe how a quark at Q^2 fragments into a "valence quark" and a number of "sea" partons of size Q_0^2. This evolution is usually characterized by the parameter ξ [see (5.32)]

$$\xi \sim \ln\left[\frac{\alpha_s(Q_0^2)}{\alpha_s(Q^2)}\right] \sim \ln\left[\frac{\ln(Q_0^2/\Lambda^2)}{\ln(Q^2/\Lambda^2)}\right]$$

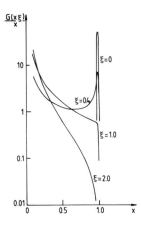

Fig.8.8. Single-parton (quark, antiquark, and gluon) inclusive distribution in a quark jet for different values of $\xi \sim \ln[\alpha_s(Q_0^2)/\alpha_s(Q)]$. The curves are calculated using an approximation to the exact QCD matrix elements [see (5.23)] /336/

Figure 8.8 shows how the parton density in a quark jet changes from $\delta(1-x)$ at $\xi = 0$ to a function decreasing with increasing x for large ξ /336/. Final state mesons are formed by recombination according to (6.17)

$$zD(z) = \int \frac{dz_q}{z_q} \frac{dz_{\bar{q}}}{z_{\bar{q}}} G(z_q, z_{\bar{q}}, Q_0^2) R(z_q, z_{\bar{q}}, z)$$

where $G(z_q, z_{\bar{q}}, Q_0^2)$ is the two-parton density within the jet. To deal with gluons, one may use the prescription /335/: "at Q_0^2 all gluons convert into quark-antiquark pairs".

Practical applications of this scheme suffer from the fact that the shape of $G(z_q, z_q, Q_0^2)$ depends crucially on the unknown ratio Q_0/Λ and that for the region of interest $Q_0 \sim 1$ GeV, the perturbative expansion starts to become unreliable.

In /335/ $Q_0/\Lambda \simeq 1.1$ is used as extracted from deep inelastic lN scattering (in a model-dependent way), and an "effective" Λ is chosen to account for higher-order corrections.

Although this procedure is somewhat questionable, the results (Fig.8.9) look encouraging, especially if vector meson production is taken into account.

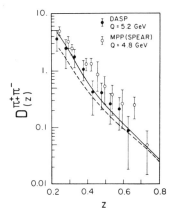

Fig.8.9. Absolute QRM predictions for quark fragmentation functions with (———) and without (- - -) vector meson contributions /335/

9. Summary

The existence of jetlike structures is shown to be a general characteristic of hadronic final states in reactions of elementary particles at high energies. The quark-parton model suggests that these jets result from the fragmentation of colored partons. Although the partons produced in various reactions may differ in their quark contents, in their color state, and in the momentum transfer Q^2 at which they are probed, the resulting jets are expected to exhibit many common features.

In the present work, properties of jets observed in

- e^+e^- annihilations into hadrons,
- hadronic decays of the T meson,
- deep inelastic lepton-nucleon scattering,
- hadron-hadron reactions with large $p_\perp$ particles, and
- normal inelastic hadronic reactions

are compared, and phenemenological approaches to describe their properties are discussed.

In the parton model, the space-time evolution of a jet may be visualized as an "inside-outside" cascade: the color charge of the active parton induces a polarization cloud which follows the parton and finally neutralizes its color.

It is shown that phenomenological models for jet fragmentation, like the Feynman-Field algorithm, can be derived from this concept.

The model of "inside-outside" cascades is further supported by QCD calculations. Here, parton fragmentation is treated as a successive branching into partons of smaller mass, yielding the same space-time structure of jets as the very naive parton model.

According to these concepts, a jet should consist of a fragmentation region and a plateau region. Parton quantum numbers are essentially retained in the fragmentation region. Jets in different environments and at different energies are related via scaling fragmentation functions, environmental independence of the fragmentation, and by the universality of the rapidity plateau.

Data on quark jets observed in e^+e^- annihilations and in lepton-nucleon interactions are consistent with these ideas. Small violations of the idealized concepts can be understood quantitatively as due to phase-space effects at low energies and QCD bremsstrahlung corrections at high energies.

In contradiction to QCD predictions, the investigation of gluon jets in T decays shows no evidence for a softening of fragmentation functions as compared to quark jets, and average multiplicities are very similar for both types of jets. However, parton fragmentation at these energies is governed by phase-space effects rather than by asymptotical parton concepts.

In hadron-hadron interactions involving large momentum transfers, pairs of jets with opposite large transverse momenta are observed. Both longitudinal and transverse properties of these jets are essentially consistent with those of "standard" quark jets. This fact, as well as measurements of inclusive particle ratios at large $p_\perp$, favor QCD models where quarks or gluons are scattered, in contrast to constituent interchange mechanisms. The absence of correlations between the charge of a pion at large $p_\perp$ and the quantum numbers of the opposite recoiling parton, proves further that parton scattering occurs through the exchange of flavor singlet fields, such as gluons.

In such interactions, where a quark is scattered out of a hadron, or more general, where a current of large Q^2 interacts with a single parton, a colored spectator system is left over and fragments. In the present work, special emphasis is put on the investigation of spectator fragments, since only here can the relation between the quark contents of a composed colored system and its fragmentation functions be studied systematically.

Fragmentation spectra of diquark spectator systems produced in νN interactions are shown to be compatible with predictions based on counting rules (DCR) or the quark recombination model (QRM). Since the shape of fragmentation spectra characterizes the quark contents of the spectator, this provides a new method to determine the type of constituent subprocess occurring, i.e., in events with large $p_\perp$ particles.

This method, having been applied to proton-proton interactions, has demonstrated that forward high-$p_\perp$ pions are fragments of valence quarks scattered at small angles.

In an analogous way fragmentation models can be used to describe hadron production at low $p_\perp$ in normal inelastic hadronic interactions, once the mechanism initiating the fragmentation is known. The absence of correlations between fast pions emitted in opposite rapidity hemispheres points towards a primary gluon exchange as compared to flavor exchange mechanisms.

Fragmentation spectra of incident hadrons into mesons and baryons are essentially consistent with the quark recombination picture and with a "pointlike" version of counting rules.

There are indications that a combination of the perturbative evolution of parton densities in jets, as given by QCD, and of a recombination mechanism provides an universal tool to understand parton fragmentation. The distribution of low "mass" partons within a jet of large Q^2 can be derived using QCD, resp. is given by the primordial parton distribution in reactions at moderate momentum transfers. The recombination principle tells how these quarks convert into hadrons.

References

1. J.D. Bjorken: Phys. Rev. *179*, 1547 (1969)
2. R.P. Feynman: Phys. Rev. Lett. *23*, 1415 (1969)
3. J.D. Bjorken, E.A. Paschos: Phys. Rev. *185*, 1975 (1969)
4. R.P. Feynman: *Photon Hadron Interactions* (W.A. Benjamin, New York 1972)
5. G. Hanson et al.: Phys. Rev. Lett. *35*, 196 (1975); *Proceedings of the 7th International Colloquium on Multiparticle Reactions* (Tutzing 1976)
6. P. Darriulat et al.: Nucl. Phys. B *107*, 429 (1976)
7. CCHK Collab., M. Della Negra et al.: Nucl. Phys. B *127*, 1 (1977)
8. B. Alper et al.: Phys. Lett. B *44*, 521 (1973)
9. M. Banner et al.: Phys. Lett. B *44*, 537 (1973)
10. F.W. Büsser et al.: Phys. Lett. B *46*, 471 (1973)
11. S.M. Berman, J.D. Bjorken, J. Kogut: Phys. Rev. D *4*, 3388 (1971)
12. D. Amati, A. Stanghellini, S. Fubini: Nuovo Cimento *26*, 896 (1962)
13. W. Ochs: Nucl. Phys. B *118*, 397 (1977)
14. K.P. Das, R.C. Hwa: Phys. Lett. B *68*, 459 (1977)
15. S.J. Brodsky, J.F. Gunion: Phys. Rev. Lett. *37*, 402 (1976)
16. B.H. Wiik, G. Wolf: *Electron-Positron Interactions*, Springer Tracts in Modern Physics, Vol. 86 (Springer Berlin, Heidelberg, New York 1979); G. Kramer et al.: to be published in Springer Tracts in Modern Physics
17. H.D. Politzer: Phys. Rev. *14*, 129 (1974)
18. T. Appelquist, H.D. Politzer: Phys. Rev. Lett. *34*, 43 (1975)
19. M. Dine, J. Sapirstein: Phys. Rev. Lett. *43*, 668 (1979)
20. DASP Collab., R. Brandelik et al.: Phys. Lett. B *70*, 125 (1977); *70*, 387 (1977); *70*, 109 (1978); *76*, 361 (1978)
21. G.P. Murtas: *Proceedings of the 19th Conference on High Energy Physics* (Tokyo 1978)
22. G. Hanson: SLAC-PUB-2118 (1978) and *Proceedings of the Conference on Gauge Theories and Leptons* (Moriond 1978)
23. PLUTO Collab., Ch. Berger et al.: Phys. Lett. B *85*, 413 (1979)
24. TASSO Collab., R. Brandelik et al.: DESY 79-61 (1979); Phys. Lett. B *89*, 243 (1979); *89*, 418 (1980); *94*, 437 (1980); *Proceedings of the 1979 International Symposium on Lepton and Photon Interactions* (Batavia 1979) p. 34
25. W. Chinowsky: Physica Scripta *19*, 65 (1979)
26. PLUTO Collab., Ch. Berger et al.: Phys. Lett. B *81*, 410 (1979); *Proceedings of the 1979 International Symposium on Lepton and Photon Interactions* (Batavia 1979) p. 19
27. JADE Collab., W. Bartel et al.: Phys. Lett. B *88*, 171 (1979)
28. S. Brand, Ch. Peyrou, R. Sosnowski, A. Wroblewski: Phys. Lett. *12*, 57 (1964)
29. H. Georgi, M. Machcek: Phys. Rev. Lett. *39*, 1237 (1977)
30. E. Farhi: Phys. Rev. Lett. *39*, 1587 (1977)
31. A. De Rújula et al.: Nucl. Phys. B *138*, 387 (1978)
32. J.D. Bjorken, S.J. Brodsky: Phys. Rev. D *1*, 1416 (1970)
33. S.D. Drell, D.J. Levy, T.M. Yan: Phys. Rev. *187*, 2159 (1969)

34 S.D. Drell, D.J. Levy, T.M. Yan: Phys. Rev. D 1, 1035 (1970); 1, 1616 (1970); 1, 2402 (1970)
35 G. Hanson: SLAC-PUB-1814 (1976) and *Proceedings of the Conference on High Energy Physics* (Tblisi 1976)
36 DASP Collab., R. Brandelik et al.: Nucl. Phys. B 148, 189 (1979)
37 Compiled by H. Wahl, CERN (private communication)
38 J. Whitmore: Phys. Rev. 10, 273 (1974)
39 D.R.O. Morrison: *Proceedings of the 4th International Conference on High Energy Collisions* (Oxford 1972)
40 M. Bardadin, L. Michejda, S. Otwinowski, R. Sosnowski: *Proceedings of the Sienna Conference on Elementary Particles*, Vol. 1 (1963) p. 628; C. Bromberg et al.: Nucl. Phys. B 107, 82 (1976)
41 PLUTO Collab., Ch. Berger et al.: Phys. Lett. B 86, 418 (1979)
42 MARK J Collab., D.P. Barber et al.: Phys. Rev. Lett. 43, 830 (1979); Phys. Lett. B 89, 139 (1979)
43 JADE Collab., W. Bartel et al.: Phys. Lett. B 91, 142 (1980)
44 J. Kogut, L. Susskind: Phys. Rev. D 9, 697 (1974)
45 J. Ellis, M.K. Gaillard, G.G. Ross: Nucl. Phys. B 111, 253 (1976); 130, 516 (1976)
46 T.A. De Grand, Y.J. Ng, S.H.H. Tye: Phys. Rev. D 16, 3251 (1977)
47 S.L. Wu, G. Zobernig: Z. Phys. C 2, 107 (1979)
48 Y.L. Dokshitzer, D.I. Dyakonov: DESY L-Trans-234 (1979)
49 G. Kramer, G. Schierholz, J. Willrodt: Phys. Lett. B 79, 249 (1978); 80, 433, (1979)
50 E. Laermann et al.: PITHA 79/14 (1979)
51 L. Van Hove: Rev. Mod. Phys. 36, 655 (1964)
52 A. Krzywicki: Nuovo Cimento 32, 1067 (1964)
53 P.P. Srivastava, G. Sudarshan: Phys. Rev. 110, 765 (1958)
54 R. Baier, J. Engels, H. Satz: Nuovo Cimento A 28, 455 (1975)
55 E.H. De Groot: Nucl. Phys. B 48, 295 (1972)
56 H. Satz, Y. Zarmi: Nuovo Cimento Lett. 15, 421 (1976)
57 W. Ernst, I. Schmitt: BI-TP-77/22
58 H. Satz: BI-TP-76/09
59 K. Kajantie, V. Karimaki: Computer Phys. Comm. 2, 207 (1971)
60 W. Kittel, L. Van Hove, W. Wojcik: Computer Phys. Comm. 1, 425 (1970)
61 J. Engels, D. Satz: Phys. Rev. D 17, 3015 (1977)
62 T. Atwood et al.: Phys. Rev. Lett. 35, 704 (1975)
63 L. Foà: Phys. Rev. 22, 1 (1975)
64 F.W. Bopp: Riv. Nuovo Cimento 1, 1 (1978) No. 8
65 K. Böckmann: BONN-HE-76-25 (1976)
66 A. Bialas, A. Kotanski: Acta Phys. Pol. B 8, 779 (1977)
67 R. Diebold: *Proceedings of the 19th International Conference on High Energy Physics* (Tokyo 1978)
68 A. Jabs: Nucl. Phys. B 34, 177 (1971)
69 G. Cocconi: Phys. Lett. B 49, 459 (1974)
70 G.I. Kopylew, M.I. Podgoretzki: Sov. Nucl. Phys. 19, 434 (1974)
71 E.H. De Groot, H. Satz: Nucl. Phys. B 130, 257 (1979)
72 K. Guettler et al.: Phys. Lett. B 64, 111 (1976)
73 R.P. Feynman: *High Energy Collisions*, ed. by C.N. Yang et al. (Gordon and Breach, New York 1969)
74 H. Fritzsch, M. Gell-Mann, H. Leutwyler: Phys. Lett B 47, 365 (1973)
75 O.W. Greenberg: Phys. Rev. Lett. 13, 598 (1964)
76 P. Hasenfratz, J. Kuti: Phys. Rev. 40, 75 (1978)
77 D.J. Gross: *Proceedings of the 19th International Conference on High Energy Physics* (Tokyo 1978) p. 486
78 J. Schwinger: Phys. Rev. 128, 2425 (1962); Theoretical Physics, Trieste Lectures, I.A.E.A. Vienna (1962)

79 A Casher, J. Kogut, L. Susskind: Phys. Rev. D *10*, 732 (1974)
80 J.D. Bjorken: *Current Induced Reactions* (Springer Berlin, Heidelberg, New York 1976)
81 B. Andersson, G. Gustafson, C. Peterson: Z. Phys. C *1*, 105 (1979); T. Sjöstrand, B. Soderberg: LU-TP-78/18; T. Sjöstrand: LU-TP-80-3
82 E. Eichten et al.: Phys. Rev. Lett. *34*, 369 (1975)
83 B. Andersson, G. Gustafson, C. Peterson: Nucl. Phys. B *136*, 27 (1978)
84 F.E. Low, K. Gottfried: Phys. Rev. D *17*, 2487 (1978)
85 A. Casher, H. Neuberger, S. Nussinov: Phys. Rev. D *20*, 179 (1979)
86 B. Andersson, G. Gustafson: Z. Phys. C *3*, 223 (1980); B. Andersson, G. Gustafson, T. Sjöstrand: LU-TP-80-1 (1980)
87 R.D. Field, R.P. Feynman: Nucl. Phys. B *136*, 1 (1978)
88 A. Krzywicki, B. Petersson: Phys. Rev. D *6*, 924 (1972)
89 J. Finkelnstein, R.D. Peccei: Phys. Rev. D *6*, 2606 (1972)
90 L.M. Seghal: *Proceedings of the International Symposium on Lepton Photon Interactions at High Energies* (1977)
91 G. Drews et al.: Phys. Rev. Lett. *41*, 1433 (1978)
92 S. Frautschi, A. Krzywicki: Z. Phys. C *1*, 43 (1979)
93 R. Hartmann: BONN-HE-78-15 (1978)
94 P.C. Bosetti et al.: Oxford 58/78 (1978)
95 W.G. Scott: Physica Scripta *19*, 179 (1979)
96 G. Altarelli: Riv. Nuovo Cimento *4*, 335 (1974)
97 J. Kogut, L. Susskind: Phys. Rev. *8*, 76 (1973)
98 G. Altarelli, G. Parisi: Nucl. Phys. B *126*, 298 (1977)
99 S.D. Drell: Phys. Rev. D *1*, 1716 (1972)
100 R. Gatto: Phys. Rev. D *7*, 2524 (1973)
101 G. Parisi: Phys. Lett. B *43*, 207 (1973); *50*, 367 (1974)
102 J. Kogut, L. Susskind: Phys. Rev. D *9*, 3391 (1974)
103 A. Casher, J. Kogut, L. Susskind: Phys. Rev. D *9*, 706 (1974)
104 K.G. Wilson: Phys. Rev. Lett. *27*, 690 (1971)
105 K.G. Wilson: Phys. Rev. B *4*, 3174 (1971)
106 A.M. Polyakov: Zh. Eksp. Teor. Fiz. *59*, 542 (1970); *60*, 1572 (1971)
107 F.J. Yndurain: Phys. Lett. B *74*, 68 (1978)
108 J. Ellis: *Proceedings of the 1979 International Symposium on Lepton and Photon Interactions* (Batavia 1979) and references quoted there
109 J.F. Owens: Phys. Lett. B *76*, 85 (1979)
110 T.A. De Grand: Nucl. Phys. B *151*, 485 (1979)
111 K. Konishi, A. Ukawa, G. Veneziano: Phys. Lett. B *78*, 243 (1978)
112 P. Cvitanovič, P. Hoyer, K. Konishi: Phys. Lett B *85*, 413 (1979)
113 A. Kirschner: Phys. Lett. B *84*, 266 (1979)
114 L. Caneschi, A. Schwimmer: Phys. Lett. B *86*, 179 (1979)
115 D. Amati, G. Veneziano: Phys. Lett. B *83*, 87 (1979)
116 G.C. Fox, S. Wolfram: CALT-68-755 (1980)
117 D. Amati: CERN TH-2650 (1979)
118 R. Odorico: Nucl. Phys. B *172*, 157 (1980)
119 H. Messel, D.F. Crawford: *Electron Photon Shower Distribution Function* (Pergamon Press Oxford 1970)
120 H. Georgi, H.D. Politzer: Nucl. Phys. B *136*, 445 (1978)
121 P.A. Rapidis et al.: Phys. Lett. B *84*, 507 (1979)
122 K. Konishi, A. Ukawa, G. Veneziano: Phys. Lett. B *80*, 259 (1979)
123 G. Sterman, S. Weinberg: Phys. Rev. Lett. *39*, 1436 (1977)
124 W. Furmanski, CERN-TH 2664 (1979)
125 B.G. Weeks: UM-HE 78-49 (1979)
126 B. Binétruy, G. Girardi: Phys. Lett. B *83*, 382 (1979)
127 I.I.Y. Bigi: Phys. Lett. B *86*, 57 (1979)
128 K. Koller, T. Walsh: DESY 78/16 (1978)
129 S.J. Brodsky: Physica Scripta *19*, 65 (1979)

130 K. Shizuya, S.H.H. Tye: Phys. Rev. Lett 41, 787 (1978)
131 M.B. Einhorn, B.G. Weeks: Nucl. Phys. B 146, 445 (1978)
132 K. Koller, T.F. Walsh, P.M. Zerwas: Phys. Lett. B 82, 263 (1979)
133 P. Hoyer et al.: DESY 78-21 (1978)
134 A. Ali et al.: DESY 79-86 (1979)
135 O. Achterberg: Frühjahrstagung DPG (Dortmund 1980)
136 TASSO Collab., R. Brandelik et al.: DESY 80/80 (1980)
137 S.J. Brodsky: SLAC-PUB-1937 (1977); *Proceedings of the Symposium on Hadron Structure and Multiparticle Production* (Kazimierz 1977)
138 G.R. Farrar, J.L. Rosner: Phys. Rev. D 10, 2226 (1974)
139 D. Sivers: Phys. Rev. D 15, 1306 (1977)
140 G.R. Farrar, J.L. Rosner: Phys. Rev. D 7, 2747 (1973)
141 R.N. Cahn, E.W. Colglazier: Phys. Rev. D 9, 2658 (1974)
142 J.L. Newmeyer, D. Sivers: Phys. Rev. D 9, 2592 (1974)
143 J.D. Bjorken: Phys. Rev. D 7, 282 (1972)
144 M. Glück, E. Reya: Nucl. Phys. B 130, 76 (1977); 145, 24 (1978)
145 G. Fontaine: *Proceedings GIF 78* (Gif sur Yvette 1978)
146 R.K. Ellis et al.: Phys. Lett. B 78, 281 (1978); CALT 68-684 (1978)
147 H.D. Politzer: Phys. Lett. B 70, 430 (1977); Nucl. Phys. B 129, 301 (1977)
148 D. Amati, R. Petronzio, G. Veneziano: CERN TH-2527 (1978)
149 N. Sakai: Phys. Lett. B 85, 67 (1979)
150 G. Altarelli et al.: CTP 793 (1979)
151 R. Baier, K. Fey: BI-TP 79/11 (1979)
152 T. Kinoshita: J. Math. Phys. 3, 650 (1962); T.D. Lee, M. Nauenberg: Phys. Rev. 133, B 1549 (1964)
153 J.H. Weis: Acta Phys. Pol. B 9, 1051 (1979)
154 L.F. Abbott, R.M. Barnett: SLAC-PUB-2325 (1979)
155 E.L. Berger: SLAC-PUB-2362 (1979)
156 A. Donnachie, P.V. Landshoff: Phys. Lett. B 95, 437 (1980)
157 O. Nachtmann: Phys. Rev. D 8, 12 (1973)
158 M. Haguenauer et al.: CERN-EP/80-1244 (1980)
159 W.S.C. Williams: *Proceedings of the International Symposium on Lepton and Photon Interactions* (Batavia 1979)
160 J. Bell et al.: Phys. Rev. D 19, 1 (1979)
161 G. Drews et al.: Phys. Rev. Lett. 41, 1433 (1978)
162 I. Cohen et al.: Phys. Rev. Lett. 40, 1614 (1978)
163 J.J. Aubert et al.: CERN-EP/80-130 (1980)
164 N. Schmitz: MPI-PAE/Exp.El. 80 (1979); 88 (1980); 89 (1980); W.G. Scott et al.: CERN-EP 79-65 (1979); CERN-EP 79-91 (1979); H. Saarikko et al.: CERN-EP 79-92 (1979)
165 J. Engels, J. Dabkowski, K. Schilling: Wu-B-79-17 (1979)
166 J.P. Berge et al.: paper submitted to Neutrino '79 Conference (Bergen 1979)
167 M. Derrick et al.: Phys. Lett. B. 88, 177 (1979); Phys. Rev. D 17, 1 (1978)
168 W.G. Scott: Physica Scripta 19, 179 (1979)
169 J.D. Bjorken, J. Kogut: Phys. Rev. D. 8, 1341 (1973)
170 P. Mazzanti, R. Odorico: Phys. Lett. B 95, 133 (1980)
171 H.M. Chan, J.E. Paton, T. Sheung-Tsun: Nucl. Phys. B 86, 479 (1975)
172 H.M. Chan et al.: Nucl. Phys. B 92, 13 (1975)
173 G. Veneziano: Phys. Lett. B 52, 220 (1974); Nucl. Phys. B 74, 365 (1974)
174 J. Dias de Deus, S. Jadach: Acta Phys. Pol. B 9, 249 (1978)
175 J.W. Chapmann et al.: Phys. Rev. D 14, 5 (1976); Phys. Rev. Lett. 36, 124 (1976)
176 C. del Papa et al.: Phys. Rev. D 13, 2934 (1976)
177 K. Bunnel et al.: Phys. Rev. D 17, 2847 (1978)
178 E.G. Gurvich: Phys. Lett. B 87, 386 (1979)
179 V.V. Ammosov et al.: Phys. Lett. B 93, 210 (1980)
180 R. Odorico: Z. Phys. C 4, 113 (1980)

181 P.M. Stevenson: ICTP/78-79/1 (1979); Nucl. Phys. B *150*, 357 (1979); *156*, 43 (1979)
182 T. Gottschalk, E. Monsay, D. Sivers: ANL-HEP-PR-79-15 (1979)
183 T. Gottschalk, D. Sivers: ANL-HEP-PR-79-07 (1979)
184 P. Binétruy, G. Girardi: Nucl. Phys. B *155*, 150 (1979)
185 D.R.O. Morrison; CERN/D.PhIII/Phys 73-46 (1973)
186 E.M. Ilgenfritz, J. Kripfganz, A. Schiller: Acta Phys. Pol. B *9*, 881 (1978)
187 K. Kinoshita, Y. Kinoshita: Phys. Lett. B *66*, 471 (1976)
188 R. Blankenbecler, S.J. Brodsky: Phys. Rev. D *10*, 2973 (1974)
189 U. Ellwanger: Nucl. Phys. B *154*, 358 (1979)
190 P.V. Landshoff: Phys. Lett. B *66*, 452 (1977)
191 P.V. Landshoff, D.M. Scott: Nucl. Phys. B *131*, 172 (1977)
192 J.F. Gunion: Phys. Lett. B *88*, 150 (1979)
193 S.J. Brodsky, J.F. Gunion: Phys. Rev. D *17*, 848 (1978)
194 H. Goldberg: Nucl. Phys. B *44*, 149 (1972)
195 M.J. Teper: RL-78-022 (1978)
196 F.E. Close: *Proceedings of the 19th International Conference on High Energy Physics* (Tokyo 1978)
197 L. Van Hove: CERN-TH-2628 (1979)
198 E. Takasugi et al.: Phys. Rev. D *20*, 211 (1979)
199 R.D. Field, R.P. Feynman: Phys. Rev. D *15*, 2590 (1977)
200 D.W. Duke, F.E. Taylor: Phys. Rev. D *17*, 1788 (1977)
201 T.A. De Grand: Phys. Rev. D *19*, 1398 (1979)
202 W. Lockmann et al.: Phys. Rev. Lett. *41*, 680 (1978)
203 R.G. Roberts, R.C. Hwa, S. Matsuda: RL-78-040 (1978)
204 L. Van Hove: Acta Phys. Pol. B *7*, 339 (1976); Nucl. Phys. B *86*, 243 (1975)
205 S. Pokorski, L. Van Hove: CERN-TH-2427 (1977)
206 J. Kalinowski, S. Pokorski, L. Van Hove: Z. Phys. C *2*, 85 (1979)
207 T.A. De Grand, H.I. Miettinen: Phys. Rev. Lett. *40*, 612 (1978)
208 J. Kuti, V.F. Weisskopf: Phys. Rev. D *4*, 3418 (1971)
209 J.F. Gunion, S.J. Brodsky, R. Blankenbecler: Phys. Lett. B *39*, 649 (1972)
210 D. Sivers, Phys. Rev. C *23*, 1 (1976)
211 V.A. Matveev, R.M. Muradyan, A.N. Tavkhelidze: Nuovo Cimento Lett. *7*, 719 (1973)
212 S.J. Brodsky, G.R. Farrar: Phys. Rev. Lett. *31*, 1153 (1973); Phys. Rev. D *11*, 1309 (1975)
213 R. Blankenbecler, S.J. Brodsky, J.F. Gunion: Phys. Lett. B *42*, 461 (1972)
214 J.F. Gunion: Phys. Rev. D *10*, 242 (1974)
215 R. Blankenbecler, S.J. Brodsky: Phys. Rev. D *10*, 2973 (1974)
216 G. Farrar: Nucl. Phys. B *77*, 429 (1974)
217 R. Blankenbecler, S.J. Brodsky, J.F. Gunion: Phys. Rev. D *12*, 3469 (1975)
218 A.I. Vainshtein, V.I. Zakharov: Phys. Lett. B *72*, 368 (1978)
219 S.J. Brodsky: quoted in Ref. 196
220 W.R. Frazer, J.F. Gunion: Phys. Rev. D *19*, 2447 (1979)
221 J.F. Gunion: Phys. Lett. B *88*, 150 (1979)
222 W.R. Innes et al.: Phys. Rev. Lett. B *66*, 286 (1978)
223 DASP II Collab., C.W. Darden et al.: Phys. Lett. B *76*, 246 (1978)
224 Ch. Berger et al.: Phys. Lett. B *76*, 243 (1978)
225 DASP II Collab., C.W. Darden et al.: Phys. Lett. B *78*, 368 (1978); Phys. Lett. B *80*, 419 (1979); DESY 80/30 (1980)
226 J.K. Bienlein et al.: Phys. Lett. B *78*, 360 (1978)
227 D. Androvs et al.: Phys. Rev. Lett. *44*, 1108 (1980); *45*, 219 (1980)
228 T. Böhringer et al.: Phys. Rev. Lett. *44*, 1111 (1980); *45*, 222 (1980)
229 H. Schröder: DESY 80/61 (1980)
230 T. Appelquist, H.D. Politzer: Phys. Rev. Lett. *34*. 43 (1975)
231 T. Appelquist, H.D. Politzer: Phys. Rev. D *12*, 1404 (1975)
232 K. Koller, H. Krasemann, T. Walsh: Z. Phys. C *1*, 71 (1979)
233 T. Appelquist, R.M. Barnett, K. Lane: Ann. Rev. Nucl. Part. Sci. *28*, 387 (1978)

234 K. Koller, T. Walsh: Phys. Lett. B. *72*, 227 (1977); *73*, 504 (1978)
235 K. Koller, T. Walsh: Nucl. Phys. B *140*, 449 (1978)
236 T.A. De Grand et al.: Phys. Rev. D *16*, 3251 (1977)
237 S.J. Brodsky et al.: Phys. Lett. B *73*, 203 (1978)
238 H. Fritzsch, K.H. Streng: Phys. Lett. B *74*, 90 (1978)
239 K. Hagiwara: Nucl. Phys. B *137*, 164 (1978)
240 K. Johnson, C.B. Thorn: Phys. Rev. D *13*, 1934 (1976)
241 A. De Rujula: *Proceedings of the 19th International Conference on High Energy Physics* (Tokyo 1978)
242 I. Montvay: Phys. Lett. B *84*, 331 (1979)
243 B. Andersson, G. Gustafson: LU-TP-79-2 (1979)
244 J. Randa: Phys. Rev. Lett. *43*, 602 (1979)
245 G. Knies: DESY 79-47 (1979)
246 F.H. Heimlich et al.: Phys. Lett. B *86*, 399 (1979)
247 W. Schmidt-Parzefall: *Proceedings of the 19th International Conference on High Energy Physics* (Tokyo 1978)
248 H. Meyer: *Proceedings of the International Symposium on Lepton and Photon Interactions* (Batavia 1979)
249 S. Brandt, H.D. Dahmen: Z. Phys. C *1*, 61 (1979)
250 A. Krzywicki et al.: Phys. Lett. B *85*, 407 (1979)
251 C. Bromberg et al.: Nucl. Phys. B *171*, 38 (1980)
252 F.W. Büsser et al.: Phys. Lett. B *51*, 306 (1974); *51*, 311 (1974)
253 B. Alper et al.: Nucl. Phys. B *100*, 237 (1975)
254 R. Kephart et al.: Phys. Rev. D *14*, 2909 (1976)
255 F.W. Büsser et al.: Nucl. Phys. B *106*, 1 (1976)
256 B. Alper et al.: Nucl. Phys. B *114*, 1 (1976); *141*, 149 (1978)
257 K. Eggert et al.: Nucl. Phys. B *98*, 49 (1973); *98*, 73 (1973); *143*, 40 (1978); Phys. Lett. B *59*, 401 (1975)
258 CCHK Collab., M. Della Negra et al.: Nucl. Phys. B *104*, 365 (1976); *127*, 1 (1977); Phys. Lett. B *59*, 401 (1975)
259 CCHK Collab., D. Drijard et al.: Nucl. Phys. B *156*, 309 (1979)
260 P. Darriulat et al.: Nucl. Phys. B *107*, 429 (1976); *110*, 365 (1976)
261 M.G. Albrow et al.: Nucl. Phys. B *135*, 461 (1978)
262 M.G. Albrow et al.: Nucl. Phys. B *145*, 305 (1978)
263 M.G. Albrow et al.: Nucl. Phys. B *160*, 1 (1980)
264 A.L.S. Angelis et al.: Phys. Lett. B *79*, 505 (1978); Physica Scripta *19*, 116 (1979); M.J. Tannenbaum: COO-2232A-83 (1979)
265 A.G. Clark et al.: Nucl. Phys. B *142*, 189 (1978); *160*, 397 (1979); Physica Scripta *19*, 79 (1979); CERN-EP/79-74 (1979)
266 R. Kourkoumelis et al.: Phys. Lett. B *83*, 257 (1979); *84*, 271 (1979); *85*, 39 (1979); *86*, 391 (1979); Nucl. Phys. B *158*, 39 (1979); Phys. Rev. Lett. *45*, 966 (1980)
267 J.W. Cronin et al.: Phys. Rev. D *11*, 3105 (1975)
268 D. Antreasyan et al.: Phys. Rev. D *19*, 764 (1979)
269 H.J. Frisch et al.: Phys. Rev. Lett. *44*, 511 (1980)
270 G. Donaldson et al.: Phys. Rev. Lett. *36*, 1110 (1976); *40*, 917 (1978)
271 C. Bromberg et al.: Nucl. Phys. B *134*, 189 (1978); *171*, 1 (1980); Phys. Rev. Lett. *42*, 1202 (1979); *43*, 561 (1979); *45*, 769 (1980)
272 M.D. Corcoran et al.: Phys. Rev. Lett. *41*, 9 (1978); Physica Scripta *19*, 95 (1979)
273 M. Dris et al.: Phys. Rev. D *19*, 1361 (1979)
274 H. Jöstlein et al.: Phys. Rev. D *20*, 53 (1979)
275 C.W. Akerloff et al.: Phys. Rev. Lett. *39*, 861 (1977)
276 D.A. Finley et al.: Phys. Rev. Lett. *42*, 1028 (1979), *42*, 1031 (1979)
277 S.D. Ellis, R. Stroynowski: Rev. Mod. Phys. *49*, 753 (1977)
278 S.D. Ellis, M. Jacob, P.V. Landshoff: Nucl. Phys. B *108*, 93 (1976)
279 M. Jacob, P.V. Landshoff: Nucl. Phys. B *113*, 395 (1976)

280 R. Cutler, D. Sivers: Phys. Rev. D *16*, 679 (1977); *17*, 196 (1978)
281 B.L. Combridge, J. Kripfganz, H. Ranft: Phys. Lett. B *70*, 234 (1977)
282 R.P. Feynman, R.D. Field, G.C. Fox: Phys. Rev. D *18*, 3320 (1978)
283 R.R. Horgan, P. Scharbach: Phys. Lett. B *81*, 215 (1979)
284 K. Hagiwara: Phys. Lett. B *84*, 241 (1979)
285 R.D. Field: CALT-68-696 (1978)
286 J.F. Owens, E. Reya, M. Glück: Phys. Rev. D *18*, 1501 (1978)
287 M.K. Chase, W.J. Stirling: Nucl. Phys. B *133*, 157 (1979)
288 R.R. Horgan, P.N. Scharbach: Phys. Lett. B *81*, 215 (1979)
289 P.V. Landshoff, J.C. Polkinghorne: Phys. Rev. D *8*, 927 (1973); *10*, 891 (1974)
290 R.P. Feynman, R.D. Field, G.C. Fox: Nucl. Phys. B *128*, 1 (1977)
291 B.L. Combridge: Phys. Rev. D *12*, 2893 (1975)
292 A.P. Contogouris, R. Gaskell, S. Papadopoulos: Phys. Rev. D *17*, 2314 (1978)
293 F. Halzen, G.A. Ringland, R.G. Roberts: Phys. Rev. Lett. *40*, 991 (1978)
294 M.K. Chase: Nucl. Phys. B *145*, 189 (1978)
295 R. Raitio, R. Sosnowski: HU-TFT-77-22 (1977)
296 Z. Kunszt, E. Pietarinen, E. Reya: Phys. Rev. D *21*, 733 (1980)
297 Z. Kunszt, E. Pietarinen: Nucl. Phys. B *164*, 45 (1980)
298 A. Schiller: J. Phys. G *5*, 1329 (1979)
299 J.B. Kogut: Phys. Lett. B *65*, 337 (1977)
300 J.F. Gunion: Phys. Rev. D *15*, 3317 (1977)
301 J.F. Owens: Phys. Rev. D *21*, 742 (1980)
302 J.F. Gunion: *The Interrelationship of the Constituent Interchange Model and QCD*, presented at the Discussion Meeting on Large Transverse Momentum Phenomena, SLAC (1978)
303 W.E. Caswell, R.R. Horgan, S.J. Brodsky: Phys. Rev. D *18*, 2415 (1978)
304 D. Jones, J.F. Gunion: Phys. Rev. D *19*, 867 (1979)
305 R. Blankenbecler, S.J. Brodsky, J.F. Gunion: Phys. Rev. D *18*, 900 (1978)
306 W. Stirling: DAMTP 79/10 (1979)
307 W. Selove: UPR-70E (1979)
308 C. Michael: *Proceedings of Workshop on Large Transverse Momentum Phenomena* (Bielefeld 1977)
309 R. Meinke: *Proceedings of Workshop on Large Transverse Momentum Phenomena* (Bielefeld 1977)
310 CCHK Collab., M. Della Negra et al.: submitted to the International Conference on High Energy Physics (Tblisi 1976)
311 R. Baier, J. Engels, B. Petersson: BI-TP 79/10 (1979)
312 J.C. Van der Velde: Physica Scipta *19*, 173 (1979)
313 W.G. Scott: *Proceedings of the International Conference on Neutrino Physics* (Purdue 1978)
314 G. Altarelli, G. Parisi, R. Petronzio: Phys. Lett. B *76*, 356 (1978)
315 M. Le Bellac: CERN 76-14 (1976)
316 CCHK Collab., D. Drijard et al.: Nucl. Phys. B *155*, 269 (1979)
317 CCHK Collab., D. Drijard et al.: Nucl. Phys. B *166*, 233 (1980)
318 H.G. Fischer, Talk given at the Third Warsaw Symposium (Jodlowy Dwor 1980)
319 J. Singh et al.: Nucl. Phys. B *140*, 189 (1978)
320 E. Reya: DESY 79/88 (1978)
321 A. Arnedo, G. Plaut: Nucl. Phys. B *107*, 262 (1976)
322 G.H. Thomas: ANL-HEP-PR-77-01 (1977)
323 J. Benecke et al.: Phys. Rev. *188*, 2159 (1969)
324 P. Capiluppi et al.: Nucl. Phys. B *79*, 189 (1974)
325 F.E. Low: Phys. Rev. D *12*, 163 (1975)
326 S. Nussinov: Phys. Rev. Lett. *34*, 1286 (1975)
327 J.F. Gunion, D.E. Soper: Phys. Rev. D *15*, 2617 (1977)
328 W.M. Morse et al.: Phys. Rev. D *15*, 66 (1977)
329 G.J. Bobbink et al.: Phys. Rev. Lett. *44*, 118 (1980)
330 M.M. Block: CERN-EP/79-82 (1979)

331 D. Cutts et al.: Phys. Rev. Lett. *43*, 319 (1979)
332 T. Edwards et al.: Phys. Rev. D *18*, 79 (1978)
333 J.R. Johnson et al.: Phys. Rev. D *17*, 1292 (1978)
334 F.E. Taylor et al.: Phys. Rev. D *14*, 1217 (1976)
335 V. Chang, R.C. Hwa: Phys. Rev. Lett. *44*, 139 (1980)
336 P. Hoyer: NORDITA-79/33 (1979)

Subject Index

Accompanying spectator 166
α_s 26, 50, 68
ALTARELLI-PARISI equations 51
Annihilation into gluons 108
Approach to scaling 19
Associated charge density 160
Away jet 129, 162
- charge 161
- charge ratio 164
- correlation with trigger 161
Away spectator 166
Azimuthal distribution of particles in jets 156

Back-to-back jet configuration 162
Back-antiback jet configuration 162
Bag model 26
Beam ratios
- high $p_\perp$ events 138
- low $p_\perp$ reactions 183
Boltzmann statistics 22
Bose-Einstein correlations 22
Bose-Einstein statistics 22
Bottom quarks 108
Branching 56
Breit frame 75
Breaking of color strings 32, 34

Cascade
- inside-outside 30
Charge
- balance 159, 177
- correlations in high $p_\perp$ events 143, 157, 177
- distribution in quark jets 41, 43, 87
- retention, see retention
Charm production in jets 60
Charge ratio
- away jet 164
- spectator jets 93, 100, 172
- trigger 139, 141
Chromodynamic Meissner effect 26
CIM, see Constituent Interchange Model
Cluster model 22
Cluster, color-singlet 57
Cms frame, in lepton-nucleon scattering 75
Coherent spectator states 74
Color 26
- center frame 110
- flux tubes 27, 31, 55, 109, 122
- octet exchange 181
- octet, fragmentation 63, 110
- octet jets 63, 110
- screening 30
- separation in QCD jets 56
- singlets 57
- strings 32, 34

- strings, tension 110
- structure, high $p_\perp$ events 121
- structure, e^+e^- annihilations 27, 110
- structure, T decays 110
- triplet exchange 181

Compensation of trigger charge 160

Constituent Interchange Model
- e^+e^- annihilations 12
- high $p_\perp$ reactions 136

Conservation of quantum numbers 159

Confinement 25, 55
- time scales 29, 56
- two-dimensional QED 27

Correlation
- function 177, 184
- high $p_\perp$ events 143, 159, 177
- quantum numbers 157, 177
- short range 22, 177
- spectator fragmentation region 177
- Uncorrelated Jet Model 15

Counting rules, see "Dimensional Counting Rules"

Cross-Section, see also "inclusive"
- e^+e^- annihilation 4
- parton-parton scattering 127

Current fragmentation region 73, 86

$D(z)$ 33, 39, 53

DCR, see Dimensional Counting Rules

Decays, T 108

Deep inelastic scattering 46, 71, 121
- color structure 73
- final states 73
- jet development 72
- proton production 94, 99
- quark-parton model 76, 79
- reference frames 75
- sea-quark scattering 73, 99

- sphericity distributions 92
- valence quark scattering 73, 99

Dimensional Counting Rules 102
- fragmentation functions 106, 173
- helicity factors 106
- low $p_\perp$ interactions 183
- pointlike coupling 107, 173
- scaling violations 106
- spin effects 106
- structure functions 105

Dipole density 28

Diquark 72, 90
- fragmentation 90, 100, 172

e^+e^- annihilations 4
- rapidity plateau 9
- scaling 8
- scaling violations 10, 67
- three-jet events 12
- total cross-section 4
- transverse momenta 7

Early scaling 21

Enhanced quark sea 97

Environmental independence 76

Event plane 12, 68

Evolution
- equations 48
- parton densities 48, 51, 55, 66, 132, 189

Exchange
- color octets 181
- color triplets 181
- quarks 181
- wee quarks 181

Experiments, high $p_\perp$ 123

Factorization 76, 79, 122, 127, 180
- experimental evidence 77
- hadron-hadron reactions 180

- violation 80
Favored fragmentation 39
Feynman-Field model 35
Flavor correlations 159, 174
Fock states 98
Fragmentation
- current 73, 86
- diquark 90, 100, 172
Fragmentation function 33, 39, 53, 82
- behavior at large x and Q^2 54
- color octet 63, 110
- Dimensional Counting Rules 106, 173
- favored 39
- Q^2 dependence 46, 52, 79
- scale breaking 47, 66, 132
- spectator 90, 100, 172
- towards/away jet in high $p_\perp$ events 148, 153
- unfavored 39
Frequency of jets, in high $p_\perp$ events 147

Gluon
- bremsstrahlung 14, 46, 62, 79, 135
- conversion into quarks 97, 190
- radiation, high $p_\perp$ events 135, 155, 168
- spin 68, 117
- structure function 50, 132
- Υ decays 108
Gluon jets 63, 110
- asymptotic behavior 64, 66
- inclusive spectra 66
- multiplicity 66, 112
- scaling violation 66

Hadron-hadron interactions 180
- factorization 180

Hard scattering expansion 137
Hard-gluon radiation
- deep inelastic lepton-nucleon reactions 79
- e^+e^- annihilations 10
- high $p_\perp$ 155, 169
Heavy-quark production 59
Helicity factors in DCR 106
Higher twists 80
- high $p_\perp$ reactions 137
High $p_\perp$ events
- charge correlations 160
- exotic trigger particles 165
- kinematics 125
- momentum balance 157
- multiplicity 153
- quantum number correlations 159
- spectator jets 129, 166
- two-particle correlations 143, 146, 160, 177
- trigger bias 128
- x_E distribution 129, 148
- z of trigger 128
High $p_\perp$ reactions
- beam ratios 138
- inclusive cross sections 128
- natural mechanisms 137
- particle ratios 138
- ϕ^3 theory 137
- QCD 130
- quark-parton model 125
High $p_\perp$ experiments 123

Inclusive cross sections
- e^+e^- annihilations 8
- jets 124, 127, 140
- high $p_\perp$ reactions 128, 133, 136
Inclusive spectra 8

- e^+e^- annihilations 8
- gluon fragments 66
- quark fragments 39
- Regge contributions 185
- spectator fragments 90, 100, 172
- uncorrelated jet model 17
- T decays 115

Inside-outside cascade 30, 38

Integral equations
- QCD jets 51
- quark fragmentation 34, 36

Isospin distribution in quark jets 43

Jet cross section 124, 127, 140

Jet development
- deep inelastic reactions 72
- high $p_\perp$ reactions 121
- integral equations 34, 36
- Lorentz invariance 33
- parton model 25, 71, 79, 125
- QCD 26, 46
- two-dimensional QED 27
- T decays 109

Jetiness 5

Jet leader 161
- charge 161

Jets 4, see also "fragmentation", "inclusive spectra", "gluon jets", "spectator jets", "Uncorrelated Jet Model"

Jets in e^+e^- annihilations 4
- limited $p_\perp$ 7, 16, 61, 82, 155, 168
- violation of scaling 12, 19, 47, 66, 132

Jets in hadron-hadron reactions 180

Jets, space-time development 27, 32, 56

Jet universality 88

Jet trigger 124

K^- production in high $p_\perp$ events 138, 165, 176

Kaon production in T decays 119

Kinematics, high $p_\perp$ events 125

KINOSHITA-LEE-NAUENBERG theorem 62

KUTI-WEISSKOPF model 98

Large $p_\perp$, see high $p_\perp$

Leading particle effect 55

Lepton-nucleon scattering 46, 71

Lepton-nucleon quasielastic scattering 75

Limiting fragmentation 180

Limited transverse momenta 7, 16, 61, 155, 168, 180

Local conservation of quantum numbers 159

Long range correlations 159

Lorentz invariance, jet formation 33

Lorentz invariant phase space 15

LOW-NUSSINOV model 181

Low $p_\perp$ phenomena 180

low $p_\perp$ interactions
- Dimensional Counting Rules 183
- Quark Recombination Model 185

LUND model 31

Mandelstam variables 125

Matrix elements
- KUTI-WEISSKOPF model 98
- Uncorrelated Jet Model 17

Master equations 51

Mass of color singlets in jets 57

Meissner effect 26

Meson wave function 95

Moments 48, 85
- nonsinglet 52
- singlet 52

Momentum balance, in high $p_\perp$ events 157

Monte-Carlo simulation of parton cascades 59

Multiplicity
- of color singlet clusters in jets 58
- deep inelastic scattering 88
- e^+e^- annihilations 4, 59
- gluon jets 65, 112
- high $p_\perp$ events 153
- quark jets 4, 59
- Υ decays 111, 114
- x and Q^2 dependence 89

Natural mechanisms 137
Nonsinglet moments 52, 85

Off-shell partons 52, 79

Particles, high $p_\perp$ 121
Particle ratios, high $p_\perp$ 138
Parton
- cascade 30, 54
- cascade, Monte-Carlo simulation 59
- density in jets 46, 55, 66, 189
- dynamical transverse momentum 157
- formfactor 56
- jets 46
- model 25, 76, 125
- model, time scales of interactions 25
- -parton cross section 127
- -parton scattering 125
- primordial transverse momentum 157
- shower 30, 54
- trajectories, classical approximation 31
- transverse momentum 134, 147, 157
Phase space 15
- scaling violations 78, 84
ϕ^3 theory, high $p_\perp$ reactions 137

Planar diagrams 57
"Pointlike" Dimensional Counting Rules 107, 173
Polarization charge 28
Pomeron model of LOW, NUSSINOV 181
P_{OUT}, definition 157
Preconfinement 55, 72
- high $p_\perp$ reactions 135
Primordial transverse momentum 135, 157
Primary mesons 36
Proton production
- deep inelastic scattering 94
- hadron-hadron reactions 182
- high $p_\perp$ reactions 171, 175, 176
Probabilistic quark model 98
Probability, for gluon bremsstrahlung 14
Pseudosphericity 116
$p^2_{\perp IN}$ 12
$p^2_{\perp OUT}$ 12

Q^2 dependence
- fragmentation functions 46, 52, 79
- structure functions 46, 79
QCD, see Quantum Chromo Dynamics
QED, two-dimensional 27
QRM, see Quark Recombination Model
Quantum Chromo Dynamics 26, 46, 50, 130
- correction of order (α_s) 80
- corrections, quark-parton model 79
- e^+e^- annihilations 67
- high $p_\perp$ reactions 130
- jets, threshold effects 60
- jets, transverse width 61
- subprocesses 130
Quantum number correlations 159
- trigger/spectator 171
Quantum numbers, local conservation 159

Quantum number structure, quark jets 43
Quantum number retention 43, 84, 180
Quark
- creation, in jets 31
- exchange 181
- fragmentation 4, 26, 39
- fragmentation function 39, 82
- fragmentation region 74
- fusion 181
- hole 74
- jets 4, 26, 39
- jets, distribution of charge 43
- jets, distribution of isospin 43
- jets, distribution of strangeness 43
- jets, extension in space-time 30
- jets, multiplicity 4, 59, 88, 153
- jets, Quark Recombination Model 189
- jets, simulation 35, 59
- jets, vector meson production 37
- model 4, 25, 76, 125
- model, QCD corrections 79
- model, deep inelastic reactions 79
- model, probabilistic 98
- -quark scattering 121
- rest frame 75
- sea 97
- sea enhancement 97
Quark Recombination Model 94
- low $p_\perp$ hadron interactions 185
Quasielastic scattering 75

R 4
Radial scaling 186
Rapidity 9
Rapidity distribution
- e^+e^- annihilations 10
- hadron-hadron interactions 182
- proton-proton reactions 10

Rapidity plateau 9, 16, 34
- gluon jets 65
- rise 20, 23
- Uncorrelated Jet Model 16
- quark jets 189
- spectator fragmentation 175
Recombination
- function 95
- model 96
- multiquark 95
- spectator fragmentation 100, 172
Recoil, parton $p_\perp$ 168
Reference frames, deep inelastic scattering 75
Regge contributions 185
Regge slope 31
Resonance production
- quark jets 34, 37
- Uncorrelated Jet Model 22
- spectator fragmentation 171
Retention
- charge 43, 87, 180
- quantum numbers 43, 84, 180
ρ meson production 22

Scale breaking 47
- fragmentation functions 52, 66, 79, 82, 132, 154
- high $p_\perp$ reactions 131, 154
- phase space effects 84
- structure functions 51, 131
Scale invariance 47
Scaling 8, 19, 77, 180
- approach to scaling 19
- inclusive spectra 8
- radial 186
- Uncorrelated Jet Model 19
- variables 75, 77

- variables, spectator fragmentation 91, 130, 168, 172
Scaling violations 12, 47, 50, 78, 86
- deep inelastic scattering 84
- dimensional counting rules 106
- fragmentation functions 52
- phase space effects 78
- Quantum Chromo Dynamics 50, 131
- Uncorrelated Jet Model 20
Scaling in x_E 129, 148
Scattering
- quark-quark 121
- parton-parton 125
Schwinger model 27, 64
Schwinger phenomenon 30
Seagull effect 10, 18, 156
Secondary mesons 36
Short-range correlations 177
- Uncorrelated Jet Model 22
Simulation of quark jets 35, 59
Singlet moments 52
Single particle trigger 124
Soft gluon radiation, high $p_\perp$ events 168
Space-time development
- parton jets 27, 32
- QCD jets 56
Spectator 72, 90
- coherence 74
- diquark 72, 100, 172
Spectator fragmentation 74, 90, 166
- azimuthal distribution in high $p_\perp$ events 168
- charge balance 177
- Dimensional Counting Rules 106, 172
- high $p_\perp$ events 172
- Quark Recombination Model 97, 175
- scaling variables 91, 130, 167, 172

- time scales 106, 176
- transverse momentum 168
- two-particle correlations 177
Spectator, quantum number correlations 171
Sphericity 5, 92
- deep inelastic scattering 92
- e^+e^- annihilations 7
- T decays 117
Spin, Dimensional Counting Rules 106
Spin, gluon 68, 117
Splitting functions 51
STERMAN-WEINBERG jets 62
Strange-meson production in jets 34, 37, 61
Strangeness distribution in quark jets 43
Strong coupling constant 26, 50, 68
Structure functions 8, 46, 127
- Dimensional Counting Rules 105
- gluon 132
- KUTI-WEISSKOPF model 98
- Q^2 dependence 46, 79
- scale breaking 51, 131
Structure of high $p_\perp$ events 141
SU(3) breaking 37
- jets 34
- quark creation 31
- sea quarks 87

Target fragmentation region 74, 86
Target mass corrections 80
Tension, color strings 110
Tests, QCD 67
Three-jet structure
- e^+e^- annihilations 12
- T decays 116
Threshold effects in jets 61
Thrust 7
Time scales
- confinement 29, 56

- parton interactions 25
- spectator fragmentation 106, 176
Towards jet 142, 151
Transverse momentum
- cut-off 7, 16, 61, 155, 168, 180
- partons in jets 61
- quark jets 38
- secondaries in high $p_\perp$ jets 155
- spectator fragments 168
- Uncorrelated Jet Model 16
Triplicity 117
Trigger bias 128
- jet trigger 128
Trigger jet 151
Tunnel effect, in color flux tubes 31
Two-dimensional QED 27
Two-particle correlations 18, 22, 159, 177, 184
- hadron-hadron interactions 184
- high $p_\perp$ events 143
- Uncorrelated Jet Model 18, 22
Two-particle density 177

UJM, see Uncorrelated Jet Model
Uncorrelated Jet Model 15, 94
- e^+e^- annihilations 20
- rapidity plateau 16
- resonance production 22
- short-range correlations 22
- two-particle correlations 18
Unfavored fragmentation 39
Universality
- jets 88
- multiplicity 88
- rapidity plateau 88
T meson 108
- annihilation into gluons 108
- decays, inclusive spectra 115
- decays, kaon production 119
- decays, matrix-element 109
- decays, sphericity axis 117
- decays, thrust 117
- decays, three-jet structure 116

Vector meson production, quark jets 37
Violation of
- factorization 80
- scaling 12, 47, 50, 78, 86
- scaling, Uncorrelated Jet Model 20

Wee parton 181
- exchange 181
Width of QCD jets 62

x distribution, gluons in T decays 109
x_E 129
x_E distribution 148
x_E scaling 129, 148
- approach to scaling 148
x_J 161
x_R 16

$<z>$, high $p_\perp$ trigger particle 128

Classified Index

Springer Tracts in Modern Physics, Volumes 36–90

This cumulative index is based upon the Physics and Astronomy Classification Scheme (PACS) developed by the American Institute of Physics

General

04 Relativity and Gravitation

Heintzmann, H., Mittelstaedt, P.: Physikalische Gesetze in beschleunigten Bezugssystemen (Vol. 47)
Stewart, J., Walker, M.: Black Holes: the Outside Story (Vol. 69)

05 Statistical Physics

Agarwal, G. S.: Quantum Statistical Theories of Spontaneous Emission and their Relation to Other Approaches (Vol. 70)
Graham, R.: Statistical Theory of Instabilities in Stationary Nonequilibrium Systems with Applications to Lasers and Nonlinear Optics (Vol. 66)
Haake, F: Statistical Treatment of Open Systems by Generalized Master Equations (Vol. 66)

07 Specific Instrumentation

Godwin, R. P.: Synchrotron Radiation as a Light Source (Vol. 51)

The Physics of Elementary Particles and Fields

11 General Theory of Fields and Particles

Brandt, R. A.: Physics on the Light Cone (Vol. 57)
Dahmen, H. D.: Local Saturation of Commutator Matrix Elements (Vol. 62)
Ferrara, S., Gatto, R., Grillo, A. F.: Conformal Algebra in Space-Time and Operator Product Expansion (Vol. 67)
Jackiw, R.: Canonical Light-Cone Commutators and Their Applications (Vol. 62)
Kundt, W.: Canonical Quantization of Gauge Invariant Field Theories (Vol. 40)
Rühl, W.: Application of Harmonic Analysis to Inelastic Electron-Proton Scattering (Vol. 57)
Symanzik, K.: Small-Distance Behaviour in Field Theory (Vol. 57)
Zimmermann, W.: Problems in Vector Meson Theories (Vol. 50)

11.30 Symmetry and Conservation Laws

Barut, A. O.: Dynamical Groups and their Currents. A Model for Strong Interactions (Vol. 50)
Ekstein, H.: Rigorous Symmetries of Elementary Particles (Vol. 37)
Gourdin, M.: Unitary Symmetry (Vol. 36)
Lopuszański, J. T.: Physical Symmetries in the Framework of Quantum Field Theory (Vol. 52)
Pauli, W.: Continuous Groups in Quantum Mechanics (Vol. 37)
Racah, G.: Group Theory and Spectroscopy (Vol. 37)
Rühl, W.: Application of Harmonic Analysis to Inelastic Electron-Proton Scattering (Vol. 57)
Wess, J.: Conformal Invariance and the Energy-Momentum Tensor (Vol. 60)
Wess, J.: Realisations of a Compact, Connected, Semisimple Lie Group (Vol. 50)

11.40 Currents and Their Properties

Furlan, G., Paver, N., Verzegnassi, C.: Low Energy Theorems and Photo- and Electroproduction Near Threshold by Current Algebra (Vol. 62)
Gatto, R.: Cabibbo Angle and $SU_2 \times SU_2$ Breaking (Vol. 53)
Genz, H.: Local Properties of σ-Terms: A Review (Vol. 61)
Kleinert, H.: Baryon Current Solving $SU(3)$ Charge-Current Algebra (Vol. 49)
Leutwyler, H.: Current Algebra and Lightlike Charges (Vol. 50)
Mendes, R. V., Ne'eman, Y.: Representations of the Local Current Algebra. A Constructional Approach (Vol. 60)
Müller, V. F.: Introduction to the Lagrangian Method (Vol. 50)
Pietschmann, H.: Introduction to the Method of Current Algebra (Vol. 50)
Pilkuhn, H.: Coupling Constants from PCAC (Vol. 55)
Pilkuhn, H.: S-Matrix Formulation of Current Algebra (Vol. 50)
Renner, B.: Current Algebra and Weak Interactions (Vol. 52)
Renner, B.: On the Problem of the Sigma Terms in Meson-Baryon Scattering. Comments on Recent Literature (Vol. 61)
Soloviev, L. D.: Symmetries and Current Algebras for Electromagnetic Interactions (Vol. 46)
Stech, B.: Nonleptonic Decays and Mass Differences of Hadrons (Vol. 50)
Stichel, P.: Current Algebra in the Framework of General Quantum Field Theory (Vol. 50)
Stichel, P.: Current Algebra and Renormalizable Field Theories (Vol. 50)
Stichel, P.: Introduction to Current Algebra (Vol. 50)
Verzegnassi, C.: Low Energy Photo- and Electroproduction, Multipole Analysis by Current Algebra Commutators (Vol. 59)
Weinstein, M.: Chiral Symmetry. An Approach to the Study of the Strong Interactions (Vol. 60)

12 Specific Theories and Interaction Models

Amaldi, E., Fubini, S. P., Furlan, G.: Electroproduction at Low Energy and Hadron Form Factors (Vol. 83)
Hofman, W.: Jets of Hadrons (Vol. 90)
Wiik, B. H., Wolf, G., Electron-Positron Interactions (Vol. 86)

12.20 Quantum Electrodynamics

Källén, G.: Radiative Corrections in Elementary Particle Physics (Vol. 46)
Olsen, H. A.: Applications of Quantum Electrodynamics (Vol. 44)

12.30 Weak Interactions

Barut, A. O.: On the S-Matrix Theory of Weak Interactions (Vol. 53)
Dosch, H. G.: The Decays of the K_0–$\bar{K}_0$ System (Vol. 52)
Gasiorowicz, S.: A Survey of the Weak Interactions (Vol. 52)
Gatto, R.: Cabibbo Angle and $SU_2 \times SU_2$ Breaking (Vol. 53)
von Gehlen, G.: Weak Interactions at High Energies (Vol. 53)
Kabir, P. K.: Questions Raised by CP-Nonconservation (Vol. 52)
Kummer, W.: Relations for Semileptonic Weak Interactions Involving Photons (Vol. 52)
Müller, V. F.: Semileptonic Decays (Vol. 52)
Paul, E.: Status of Interference Experiments with Neutral Kaons (Vol. 79)
Pietschmann, H.: Weak Interactions at Small Distances (Vol. 52)
Primakoff, H.: Weak Interactions in Nuclear Physics (Vol. 53)
Renner, B.: Current Algebra and Weak Interactions (Vol. 52)
Riazuddin: Radiative Corrections to Weak Decays Involving Leptons (Vol. 52)
Rothleitner, J.: Radiative Corrections to Weak Interactions (Vol. 52)
Segre, G.: Unconventional Models of Weak Interactions (Vol. 52)
Stech, B.: Non Leptonic Decays (Vol. 52)

12.40 Strong Interactions, Regge Polee Theory, Dual Models

Ademollo, M.: Current Amplitudes in Dual Resonance Models (Vol. 59)
Chung-I Tan: High Energy Inclusive Processes (Vol. 60)
Collins, P. D. B.: How Important Are Regge Cuts? (Vol. 60)
Collins, P. D. B., Gault, F. D.: The Eikonal Model for Regge Cuts in Pion-Nucleon Scattering (Vol. 63)
Collins, P. D. B., Squires, E. J.: Regge Poles in Particle Physics (Vol. 45)
Contogouris, A. P.: Certain Problems of Two-Body Reactions with Spin (Vol. 57)
Contogouris, A. P.: Regge Analysis and Dual Absorptive Model (Vol. 63)
Dietz, K.: Dual Quark Models (Vol. 60)
van Hove, L.: Theory of Strong Interactions of Elementary Particles in the GeV Region (Vol. 39)
Huang, K.: Deep Inelastic Hadronic Scattering in Dual-Resonance Model (Vol. 62)
Landshoff, P. V.: Duality in Deep Inelastic Electroproduction (Vol. 62)
Michael, C.: Regge Residues (Vol. 55)
Oehme, R.: Complex Angular Momentum (Vol. 57)
Oehme, R.: Duality and Regge Theory (Vol. 57)
Oehme, R.: Rising Cross-Sections (Vol. 61)
Rubinstein, H. R.: Duality for Real and Virtual Photons (Vol. 62)
Rubinstein, H. R.: Physical N-Pion Functions (Vol. 57)
Satz, H.: An Introduction to Dual Resonance Models in Multiparticle Physics (Vol. 57)
Schrempp-Otto, B., Schrempp, F.: Are Regge Cuts Still Worthwhile? (Vol. 61)
Squires, E. J.: Regge-Pole Phenomenology (Vol. 57)

13.40 Electromagnetic Properties of Hadrons

Buchanan, C. D., Collard, H., Crannell, C., Frosch, R., Griffy, T. A., Hofstadter, R., Hughes, E. B., Nöldeke, G. K. Oakes, R. J., Van Oostrum, K. J., Rand, R. E., Suelzle, L., Yearian, M. R., Clark, B., Herman, R., Ravenhall, D. G.: Recent High Energy Electron Investigations at Stanford University (Vol. 39)
Gatto, R.: Theoretical Aspects of Colliding Beam Experiments (Vol. 39)
Gourdin, M.: Vector Mesons in Electromagnetic Interactions (Vol. 55)
Huang, K.: Duality and the Pion Electromagnetic Form Factor (Vol. 62)
Wilson, R.: Review of Nucleon Form Factors (Vol. 39)

13.60 Photon and Lepton Interactions with Hadrons

Brinkmann, P.: Polarization of Recoil Nucleons from Single Pion Photoproduction. Experimental Methods and Results (Vol. 61)
Donnachie, A.: Exotic Electromagnetic Currents (Vol. 63)
Drees, J.: Deep Inelastic Electron-Nucleon Scattering (Vol. 60)
Drell, S. D.: Special Models and Predictions for Photoproduction above 1 GeV (Vol. 39)
Fischer, H.: Experimental Data on Photoproduction of Pseudoscalar Mesons at Intermediate Energies (Vol. 59)
Foà, L.: Meson Photoproduction on Nuclei (Vol. 59)
Frøyland, J.: High Energy Photoproduction of Pseudoscalar Mesons (Vol. 63)
Furlan, G., Paver, N., Verzegnassi, C.: Low Energy Theorems and Photo- and Electroproduction Near Threshold by Current Algebra (Vol. 62)
von Gehlen, G.: Pion Electroproduction in the Low-Energy Region (Vol. 59)
Heinloth, K.: Experiments on Electroproduction in High Energy Physics (Vol. 65)
Höhler, G.: Special Models and Predictions for Pion Photoproduction (Low Energies) (Vol. 39)
von Holtey, G.: Pion Photoproduction on Nucleons in the First Resonance Region (Vol. 59)
Landshoff, P. V.: Duality in Deep Inelastic Electroproduction (Vol. 62)

Llewellyn Smith, C. H.: Parton Models of Inelastic Lepton Scattering (Vol. 62)
Lücke, D., Söding,P.: Multipole Pion Photoproduction in the s-Channel Resonance Region (Vol. 59)
Osborne, L. S.: Photoproduction of Mesons in the GeV Range (Vol. 39)
Pfeil, W., Schwela, D.: Coupling Parameters of Pseudoscalar Meson Photoproduction on Nucleons (Vol. 55)
Renard, F. M.: ϱ-ω Mixing (Vol. 63)
Rittenberg, V.: Scaling in Deep Inelastic Scattering with Fixed Final States (Vol. 62)
Llewellyn Smith, C. H.: Parton Models of Inelastic Lepton Scattering (Vol. 62)
Lüke, D., Söding, P.: Multipole Pion Photoproduction in the s-Channel Resonance Region (Vol. 59)
Rollnik, H., Stichel, P.: Compton Scattering (Vol. 79)
Rubinstein, H. R.: Duality for Real and Virtual Photons (Vol. 62)
Rühl, W.: Application of Harmonic Analysis to Inelastic Electron-Proton Scattering (Vol. 57)
Schildknecht, D.: Vector Meson Dominance, Photo- and Electroproduction from Nucleons (Vol. 63)
Schilling, K.: Some Aspects of Vector Meson Photoproduction on Protons (Vol. 63)
Schwela, D.: Pion Photoproduction in the Region of the Δ (1230) Resonance (Vol. 59)
Wolf, G.: Photoproduction of Vector Mesons (Vol. 57)

13.75 Hadron-Induced Reactions

Atkinson, D.: Some Consequences of Unitary and Crossing. Existence and Asymptotic Theorems (Vol. 57)
Basdevant, J. L.,: $\pi\pi$ Theories (Vol. 61)
DeSwart, J. J., Nagels, M. M., Rijken, T. A., Verhoeven, P.A.: Hyperon-Nucleon Interaction (Vol. 60)
Ebel, G., Julius, D., Kramer, G., Martin, B. R., Müllensiefen, A., Oades, G., Pilkuhn, H., Pišút, J., Roos, M., Schierholz, G., Schmidt, W., Steiner, F., DeSwart, J. J.: Compilating of Coupling Constants and Low-Energy Parameters (Vol. 55)
Gustafson, G., Hamilton, J.: The Dynamics of Some π-N Resonances (Vol. 57)
Hamilton, J.: New Methods in the Analysis of πN Scattering (Vol. 57)
Kramer, G.: Nucleon-Nucleon Interactions below 1 GeV/c (Vol. 55)
Lichtenberg, D. B.: Meson and Baryon Spectroscopy (Vol. 36)
Martin, A. D.: The ΛKN Coupling and Extrapolation below the KN Threshold (Vol. 55)
Martin, B. R.: Kaon-Nucleon Interactions below 1 GEV/c (Vol. 55)
Morgan, D., Pišút, J.: Low Energy Pion-Pion Scattering (Vol. 55)
Oades, G. C.: Coulomb Corrections in the Analysis of πN Experimental Scattering Data (Vol. 55)
Pišút, J.: Analytic Extrapolations and the Determination of Pion-Pion Shifts (Vol. 55)
Wanders, G.: Analyticity, Unitary and Crossing-Symmetry Constraints for Pion-Pion Partial Wave Amplitudes (Vol. 57)
Zinn-Justin, J.: Course on Padé Approximants (Vol. 57)

Nuclear Physics

21 Nuclear Structure

Arenhövel, H., Weber, H. J.: Nuclear Isobar Configurations (Vol. 65)
Cannata, F., Überall, H.: Giant Resonance Phenomena in Intermediate-Energy Nuclear Reactions (Vol. 89)
Racah, G.: Group Theory and Spectroscopy (Vol. 37)
Singer, P.: Emission of Particles Following Muon Capture in Intermediate and Heavy Nuclei (Vol. 71)
Überall, H.: Study of Nuclear Structure by Muon Capture (Vol. 71)
Wildermuth, K., McClure, W.: Cluster Representations of Nuclei (Vol. 41)

21.10 Nuclear Moments

Donner, W., Süßmann, G.: Paramagnetische Felder am Kernort (Vol. 37)
Zu Putlitz, G.: Determination of Nuclear Moments with Optical Double Resonance (Vol. 37)
Schmid, D.: Nuclear Magnetic Double Resonance – Principles and Applications in Solid State Physics (Vol. 68)

21.30 Nuclear Forces and

21.40 Few Nucleon Systems

DeSwart, J. J., Nagels, M. M., Rijken, T. A., Verhoeven, P.A.: Hyperon-Nucleon Interactions (Vol. 60)
Kramer, G.: Nucleon-Nucleon Interactions below 1 GeV/c (Vol. 55)
Levinger, J. S.: The Two and Three Body Problem (Vol. 71)

23 Weak Interactions

Gasiorowicz, S.: A Survey of the Weak Interaction (Vol. 52)
Primakoff, H.: Weak Interactions in Nuclear Physics (Vol. 53)

25.30 Lepton-Induced Reactions and Scattering

Theißen, H.: Spectroscopy of Light Nuclei by Low Energy (70 MeV) Inelastic Electron Scattering (Vol. 65)
Überall, H.: Electron Scattering, Photoexcitation and Nuclear Models (Vol. 49)

28.20 Neutron Physics

Koester, L.: Neutron Scattering Lengths and Fundamental Neutron Interactions (Vol. 80)
Springer, T.: Quasi-Elastic Scattering of Neutrons for the Investigation of Diffusive Motions in Solids and Liquids (Vol. 64)
Steyerl, A.: Very Low Energy Neutrons (Vol. 80)

29 Experimental Methods

Panofsky, W. K. H.: Experimental Techniques (Vol. 39)
Strauch, K.: The Use of Bubble Chambers and Spark Chambers at Electron Accelerators (Vol. 39)

Atomic and Molecular Physics

31 Electronic Structure of Atoms and Molecules, Theory

Donner, W., Süßmann, G.: Paramagnetische Felder am Kernort (Vol. 37)

32 Atomic Spectra and Interactions with Photons

Racah, G.: Group Theory and Spectroscopy (Vol.37)
Zu Putlitz, G.: Determination and Nuclear Moments with Optical Double Resonance (Vol. 37)

34 Atomic and Molecular Collision Processes and Interactions

Dettmann, K.: High Energy Treatment of Atomic Collisions (Vol. 58)
Langbein, D.: Theory of Van der Waals Attraction (Vol. 72)
Seiwert, T.: Unelastische Stöße zwischen angeregten und unangeregten Atomen (Vol. 47)

Classical Fields of Phenomenology

41.70 Particles in Electromagnetic Fields

Olson, C. L.: Collective Ion Acceleration with Linear Electron Beams (Vol. 84)
Schumacher, U.: Collective Ion Acceleration with Electron Rings (Vol.84)

41.80 Particle Optics

Hawkes, P. W.: Quadrupole Optics (Vol. 42)

42 Optics

42.50 Quantum Optics

Agarwal, G. S.: Quantum Statistical Theories of Spontaneous Emission and their Relation to Other Approaches (Vol. 70)
Graham, R.: Statistical Theory of Instabilities in Stationary Nonequilibrium Systems with Applications to Lasers and Nonlinear Optics (Vol. 66)
Haake, F.: Statistical Treatment of Open Systems by Generalized Master Equations (Vol. 66)
Schwabl, F., Thirring, W.: Quantum Theory of Laser Radiation (Vol. 36)

42.72 Optical Sources

Godwin, R. P.: Synchrotron Radiation as a Light Source (Vol. 51)

Fluids, Plasmas

51 Kinetics and Transport Theory of Fluids; Physical Properties of Gases

Geiger, W., Hornberger, H., Schramm K.-H.: Zustand der Materie unter sehr hohen Drücken und Temperaturen (Vol. 46)
Hess, S.: Depolarisierte Rayleigh-Streuung und Strömungsdoppelbrechung in Gasen (Vol. 54)

Condensed Matter, Mechanical and Thermal Properties

61 Structure of Liquids and Solids

Behringer, J.: Factor Group Analysis Revisited and Unified (Vol. 68)
Dederichs, P. H., Zeller, R.: Dynamical Properties of Point Defects in Metals (Vol. 87)
Lacmann, R.: Die Gleichgewichtsform von Kristallen und die Keimbildungsarbeit bei der Kristallisation (Vol. 44)
Langbein, D.: Theory of Van der Waals Attraction (Vol. 72)
Leibfried, G., Breuer, N.: Point Defects in Metals I: Introduction to the Theory (Vol. 81)
Schroeder, K.: Theory of Diffusion Controlled Reactions of Point Defects in Metals (Vol. 87)
Springer, T.: Quasi-elastic Scattering of Neutrons for the Investigation of Diffusive Motions in Solids and Liquids (Vol. 64)
Steeb, S.: Evaluation of Atomic Distribution in Liquid Metals and Alloys by Means of X-Ray. Neutron and Electron Diffraction (Vol. 47)

62 Mechanical and Acoustical Properties and

63 Lattice Dynamics

Ludwig, W.: Recent Developments in Lattice Theory (Vol. 43)
Schramm, K.-H.: Dynamisches Verhalten von Metallen unter Stoßwellenbelastung (Vol. 53)

Condensed Matter: Electronic Structure, Electrical, Magnetic and Optical Properties

71 Electron States and

72 Electronic Transport in Condensed Matter

Bauer, G.: Determination of Electron Temperatures and of Hot-Electron Distribution Functions in Semiconductors (Vol. 74)
Bennemann, K.H.: A New Self-consistent Treatment of Electrons in Crystals (Vol. 38)
Daniels, J.: v. Festenberg, C., Raether, H., Zeppenfeld, K.: Optical Constants of Solids by Electron Spectroscopy (Vol. 54)
Dornhaus, R., Nimtz, G.: The Properties and Applications of $Hg_{1-x}Cd_xTe$ Alloy Systems (Vol. 78)
Feitknecht, J.: Silicon Carbide as a Semiconductor (Vol. 58)